Contents

Introduction

As their name indicates, Free-Standing Mathematics Units, or FSMU, are mathematical courses separate from any other qualification. They have evolved from the Government's decision to encourage more students post-16 to continue with a numerical source.

The units are at three levels:

- Foundation level, broadly equivalent to the lower grades of GCSE,
- Intermediate Level, broadly equivalent to the upper levels of GCSE, and
- Advanced Level, equivalent to the Mathematics found in GCE Advanced or Advanced Supplementary

The units are designed to complement the Key Skills which are used in the Application of Number and to incorporate the majority of these key skills in an integrated post-16 course.

This book is written for students following either:

- Managing Money at Foundation level or
- Calculating Finance at Intermediate Level.

These units cover topics which are comparable to those studied in the Money Management module of the SEG Modular GCSE course. These financial topics are also an ideal study for those following a GNVQ course of a financial nature – for example, Business and Finance – and also for those students following a non-vocational course in a similar area, such as GCE Advanced or Advanced Subsidiary Business Studies.

Candidates studying Managing Money should use the grid opposite, which gives the FSMU level reflected in each topic. In addition to preparing the student for the written examination, the topics contained in this book will enable the student to have the knowledge necessary to complete the portfolio which forms fifty per cent of the final assessment.

Students using textbooks sometimes worry if the solution at the end of the book differs slightly from their own answer. Very often this is caused by a misunderstanding about accuracy. In this book answers involving money should be given to the nearest penny and exact answers should always be given where possible. Where appropriate, other levels of accuracy can be required. As a guide, students should use at least four figures in their working and give answers to three significant figures. The answer must be rounded to the third figure.

We are grateful to the HMSO for information from the *Annual Abstract of Statistics*, and to the Southern Examining Group for permission to reproduce questions from past GCSE Modular Mathematics examination papers.

We hope that this book will be of great benefit to lecturers and students alike. If you have any suggestions for enhancing its usefulness in future editions, please contact us via the Oxford office of OUP.

Brian Gaulter and Leslye Buchanan
Hampshire, 2000

Free-Standing Mathematics Units

DATA

Brian Gaulter +
Leslye Buchanan

OXFORD
UNIVERSITY PRESS

OXFORD
UNIVERSITY PRESS

Great Clarendon Street, Oxford OX2 6DP

Oxford University Press is a department of the University
of Oxford. It furthers the University's objective of excellence in
research, scholarship, and education by publishing worldwide in

Oxford New York
Athens Auckland Bangkok Bogotá Buenos Aires
Calcutta Cape Town Chennai Dar es Salaam
Delhi Florence Hong Kong Istanbul Karachi
Kuala Lumpur Madrid Melbourne Mexico City
Mumbai Nairobi Paris São Paulo Shanghai
Singapore Taipei Tokyo Toronto Warsaw

with associated companies in
Berlin Ibadan

Oxford is a registered trade mark of Oxford University Press
in the UK and in certain other countries

© Oxford University Press 2000

First published 2000

All rights reserved. No part of this publication may be reproduced,
stored in a retrieval system, or transmitted, in any form or by any
means, without prior permission in writing of Oxford University
Press, or as expressly permitted by law, or under terms agreed with
the appropriate reprographics rights organisation. Enquiries
concerning reproduction outside the scope of the above should be
sent to the Rights Department, Oxford University Press, at the
above.

You must not circulate this book in any other binding or cover and
you must impose this same condition on any acquiror.

British Library Cataloguing in Publication Data

Data available

ISBN 0 19 914798 1

Typeset and illustrated by Tech-Set Limited

Printed and bound in Great Britain

FSMU LEVELS MATRIX

Book Section	Foundation	Intermediate
1.1	✓	
1.2	✓	
1.3		✓
2.1	✓	
2.2	✓	
2.3	✓	
2.4	✓	
3.1	✓	
3.2	✓	
3.3	✓	
3.4	✓	
4.1	✓	
4.2	✓	
4.3	✓	
4.4	✓	
4.5	✓	
5	✓	
6.1		✓
6.2		✓
6.3		✓
6.4		✓
6.5		✓
6.6		✓
6.7		✓
6.8		✓
7.1	✓	
7.2	✓	
7.3	✓	
7.4	✓	

Book Section	Foundation	Intermediate
8.1	✓	
8.2	✓	
8.3	✓	
8.4	✓	
8.5		✓
8.6	✓	
8.7		✓
8.8		✓
8.9	✓	
8.10		✓
8.11		✓
9.1	✓	
9.2	✓	
9.3	✓	
9.4	✓	
9.5	✓	
9.6	✓	
10.1		✓
10.2		✓
10.3		✓
10.4		✓
10.5		✓
11.1	✓	
11.2	✓	
11.3	✓	
12.1	✓	
12.2	✓	
12.3		✓
12.4		✓
13.1		✓
13.2		✓

01 Basic Numeracy

Even in prehistoric times the basis of arithmetic was being developed. The earliest number system was one, two, many. When races began to settle and become farmers, craftsmen and traders, there was a need for more sophisticated number systems for counting, recording and calculation. As different civilisations evolved, they developed different number systems. Our system is based on the number 10. Other civilisations used different systems based on the number 60 (Babylonia), 20 (Mayan) or 5. The Romans used letters to represent numbers (e.g. I, V, X, C for 1, 5, 10, 100).

When trade and communication spread beyond the local area to the rest of the country and abroad, it became essential to have a common system of numbers which was efficient.

The **decimal** system, based on the number 10, was eventually accepted as the most convenient. Its advantages are that any number can be written using only ten symbols, 0, 1, 2, 3, 4, 5, 6, 7, 8, 9 (called **digits**) and **place value**.

1.1 Integers and decimal fractions

Place value

An abacus is a frame with beads sliding on wires, which was used as a counting aid before the adoption of the ten digits. It is still used for this purpose in parts of Asia.

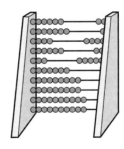

Beads on the first wire have the same value of 1 unit, but a bead on the second wire has a value which is the total of the beads on the first wire, i.e. 10. Each bead on the third wire then has the value $10 \times 10 = 100$, and so on.

The **decimal system** works in the same way, but with digits written in columns instead of beads on wires. Working from right to left, each column is worth 10 times the previous column:

$$15 = 5 \times \mathbf{1} + 1 \times \mathbf{10} = \text{fifteen}$$
$$352 = 2 \times \mathbf{1} + 5 \times \mathbf{10} + 3 \times \mathbf{100} = \text{three hundred and fifty two}$$
$$1234 = 4 \times \mathbf{1} + 3 \times \mathbf{10} + 2 \times \mathbf{100} + 1 \times \mathbf{1000}$$
$$= \text{one thousand, two hundred and thirty four}$$

If there is no digit in a particular column, the place is filled by **0**, which is called a **place-holder**.

$$105 = 5 \times \mathbf{1} + 1 \times \mathbf{100} = \text{one hundred and five}$$
$$150 = 5 \times \mathbf{10} + 1 \times \mathbf{100} = \text{one hundred and fifty}$$

Note. Whole numbers are also called **integers**.

Similarly, working from left to right, each column is worth one tenth of the previous column.

Each number in the column to the right of the units is worth $1 \div 10$, i.e. one tenth, which is a **decimal fraction**.

To separate whole numbers from decimal fractions a **decimal point** is used:

$0.1 = \frac{1}{10}$ = one tenth

$0.01 = \frac{1}{100}$ = one hundredth

$0.001 = \frac{1}{1000}$ = one thousandth

Multiplication and division by 10, 100, 1000, etc., is very easy in the decimal system.

EXAMPLE 1

Multiply 34 by: **a** 10   **b** 1000

a When a number is multiplied by 10, each digit becomes worth 10 times more, i.e. moves **1 column** to the **left**.
The empty space in the units column is filled by a zero:
$34 \times 10 = 340$

b When a number is multiplied by 1000, each digit becomes worth $10 \times 10 \times 10$ more, i.e. moves **3 columns** to the **left**.
The three empty spaces in the units, tens and hundreds columns are filled by zeros:
$34 \times 1000 = 34000$

With numbers of more than four digits, it is usual to leave a small gap after each group of three digits, e.g. 34 000.

EXAMPLE 2

Divide 2030 by: **a** 10   **b** 100

a When a number is divided by 10, each digit becomes worth 10 times less, i.e. moves **1 column** to the **right**.
The zero in the units column is no longer needed:
$2030 \div 10 = 203$

b When a number is divided by 100, each digit is worth 10×10 times less, i.e. moves **2 columns** to the **right**.

The 3 in the tens column will move into the **tenths column**:
$2030 \div 100 = 20.30$ or 20.3

EXERCISE 1.1

1 What value do the following digits have in the given number?

    a 5 in 250   **d** 4 in 624

    b 7 in 1725   **e** 6 in 12.6

    c 3 in 3012   **f** 2 in 34.529

2 Write the following numbers in order of size, smallest first:

    a 54   17   45   10   21   86

    b 104   23   14   230   32   203

    c 60   6   600   61   610   601

    d 99   101   11   110   1001   999

    e 399   400   300   297   420

3 What is the largest number which can be made from the digits 1, 7, 5 and 0?

4 What is the smallest number which can be made from the digits 3, 9, 6 and 2?

5 Find the answers to the following multiplications:

   a 43×10 **d** 13.2×10 **g** 5.16×100

   b 167×100 **e** 5.8×100 **h** 7.03×1000

   c 20×100 **f** 93.58×10 **i** 0.13×1000

6 Find the answers to the following divisions:

   a $320 \div 10$ **d** $4030 \div 100$ **g** $2.7 \div 1000$

   b $300 \div 100$ **e** $65 \div 100$ **h** $0.51 \div 10$

   c $4000 \div 1000$ **f** $65 \div 1000$ **i** $0.02 \div 100$

7 What number is formed when:

   a 1 is added to 99   **c** 1 is added to 4099

   b 2 is added to 398   **d** 2 is added to 6198?

Decimals and the four rules

The four rules of arithmetic are **addition**, **subtraction**, **multiplication** and **division**. It is generally easiest to perform calculations involving decimals on a calculator (see Unit 2). However, in simple cases, you should also be able to find an answer without the aid of a calculator.

EXAMPLE 1

a Add 34.5, 9.7 and 56.12.

b Subtract 91.72 from 164.6.

First, write the numbers in a column so that the decimal points are in line, then each digit is in its correct place.

a   34.5

      9.7

     56.12

   100.32

b 164.60

    91.72

    72.88

EXAMPLE 2

a Multiply 271.3 by 9.

b Divide 384.8 by 7.

a   271.3

        9

   2441.7

b   54.9

  7)384.3

EXERCISE 1.2

Calculate, without the aid of a calculator:

1 $36.2 + 5.7$

2 $104.9 + 75.4$

3 $6.7 + 51.09 + 76.18$

4 $72.9 - 56.2$

5 $89.13 - 72.42$

6 $121.6 - 69.85$

7 24.2×7

8 78.04×9

9 147.5×12

10 $54.6 \div 7$

11 $650.79 \div 9$

12 $526.46 \div 11$

Combining the four operations

Does $3 + 4 \times 2 = 7 \times 2 = 4$
or
does $3 + 4 \times 2 = 3 + 8 = 11$?

When more than one **operation** is used in a calculation there has to be an agreed order for combining the numbers.

The order used in the calculation is:

(i) brackets
(ii) multiplications and divisions
(iii) additions and subtractions

$\therefore \quad 3 + 4 \times 2 = 3 + 8 = 11$

EXAMPLE

Find: **a** $7 + 3 \times 6 - 1$ **b** $(7 + 3) \times 6 - 1$ **c** $7 + 3 \times (6 - 1)$

a $7 + 3 \times 6 - 1 = 7 + 18 - 1 = 24$

b $(7 + 3) \times 6 - 1 = 10 \times 6 - 1 = 60 - 1 = 59$

c $7 + 3 \times (6 - 1) = 7 + 3 \times 5 = 7 + 15 = 22$

EXERCISE 1.3

The answers to questions 1–10 should be found without the aid of a calculator. (You may use a calculator to check your answers.)

1	$5 + 7 \times 3$	**6**	$(21 + 17 - 10) \div 4$
2	$(5 + 7) \times 3$	**7**	$16 \div 4 - 20 \div 5$
3	$10 \div (5 - 3)$	**8**	$528 \div (150 - 142)$
4	$10 \div 5 - 1$	**9**	$150 - 90 \div 45 - 15$
5	$11 - 15 \div 3 \times 2$	**10**	$(150 - 90) \div (45 - 15)$

11 An artist buys 9 paint brushes costing 56p each.
How much change will be received from a £10 note?

12 Canvas costs £3.50 per square metre.
How many square metres can be bought for £14?

13 An office buys two computers for £889.08 each (including VAT) and a laser printer costing £1643.83. The budget for this expenditure is £3500. How much money remains after the purchases?

14 A theatre can seat 564 people in the stalls, 228 in the circle, and 196 in the balcony.

a How many people can the theatre seat in total?

Seats in the stalls cost £4.50, circle seats cost £6.50, and balcony seats cost £3.75.
b How much will the theatre take in ticket sales if it has a full house?

15   A freelance typist works at 60 wpm. She charges 0.25p per word.

    a How long will she take to type a document which is 7620 words long?

    b How much will she receive for this document?

16   Packs of 40 nappies cost £6.95.
How many packs can be bought for £50?
How much change will be received?

17   Mrs Halliday uses her own estate car when she delivers meals on wheels and keeps a record of her mileage.

Week of 24/1/00	Mon	Wed	Fri	Sun
Mileage	12.3	11.5	13.9	9.8

For the week shown above:

    a what was her total mileage for the week?

    b how much does she claim for the week if she claims 27.2p per mile?

18   The diagram shows an extract from a holiday brochure:

Hotel	Golden Sands		Park Royal		Ocean Lodge	
Dates	14 days	21 days	14 days	21 days	14 days	21 days
Mar 2–Mar 29	323	368	337	382	284	321
Mar 30–Apr 26	373	425	387	439	334	377
Apr 27–May 24	338	385	352	400	299	339
May 25–Jun 21	367	418	382	434	329	372
Jun 22–Jul 19	399	455	414	470	361	408
Single room supp.	£2.30 per day		£2.70 per day		£2.90 per day	

Find the cost of a holiday for three adults staying at the Park Royal for 14 days from May 25. The third adult will require a single room.

19   It costs £44.50 per day to hire a car plus £0.06 per mile travelled.
How much does it cost to hire a car for 3 days to travel 450 miles?

20   A foundry makes accessories for fireplaces. A pair of brass fire dogs weighs 1.7 kg, a set of fire irons weighs 2.04 kg, and a fire screen weighs 3.65 kg.

    a What is the total weight of the accessories for one fireplace?

A van carries a load of up to 1000 kg.

    b How many sets of the above accessories can the van carry?

21   A machine buts lengths of hollow metal rod for lamps. Each section is 20.6 cm long.

    a How many sections can be cut from a 200 cm length of rod?

    b What length of rod will be wasted?

1.2 Directed numbers

The negative sign has two distinct uses in mathematics:

(i) as a **subtraction** operation, e.g. $6 - 4 = 2$,
(ii) as a **direction** symbol, e.g. $-7\,°C$.

If we wish to show a temperature which is $7\,°C$ *below* zero, we can write $^-7\,°C$ *or* $-7\,°C$.

If a car travels 20 miles in one direction and then 15 miles in the reverse direction, we can write the distances travelled as $^+$**20 miles** and $^-$**15 miles**.

The numbers $^-7$, $^+20$ and $^-15$ are called **directed numbers**.

On your calculator you will see that there are two keys with negative symbols:

$\boxed{-}$ for subtraction

$\boxed{+/-}$ for direction.

Directed numbers can be represented on a horizontal or a vertical number line.

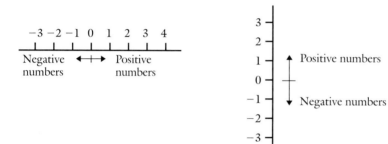

Addition and subtraction

Think of additions and subtractions of numbers as movements *up* and *down* the vertical number line.

Start at zero, move up 7 and then down 3.
Your position is now 4 *above* zero.

$$\text{i.e.} \quad 7 + (-3) = +4$$

Subtracting two numbers is the same as finding the difference between them:

$$5 - (-3)$$

is the difference between being 3 *below* zero and being 5 *above* zero on the number line.

From 3 below to 5 above you must move up 8. Therefore

$$5 - (-3) = +8 \ \textit{or} \ 5 - (-3) = 5 + 3 = 8$$

(Note that $-(-3) = +3$.)

$$-4 - (-2)$$

is the difference between being 2 below zero and being 4 below zero.

From 2 *below* to 4 *below* you must move DOWN 2, i.e.

$$-4 - (-2) = -2$$

or use the rule $- \times - = +$ to give

$$-4 - (-2) = -4 + 2 = -2$$

EXERCISE 1.4

Find the answers to the following, without the aid of a calculator.

1   a $-5 + 7$ c $-2 + -8$ e $-21 + 27$ g $-19 + 19$

    b $3 + (-7)$ d $16 + (-12)$ f $35 + 5$ h $-35 + (-14)$

2   a $7 - 8$ c $-9 - 4$ e $-10 + (-7)$ g $25 - (-25)$

    b $12 - (-5)$ d $11 - (-9)$ f $-15 - (-15)$ h $100 - 64.$

Multiplication and division

Going *up* 2 three times would mean you were now 6 *above* zero. So

$$(+2) \times 3 = +6$$

$$\text{or} \quad 2 \times 3 = 6 \qquad \text{also} \quad 3 \times 2 = 6$$

Going *down* 2 three times would mean you were now 6 *below* zero. So

$$(-2) \times 3 = -6$$

$$\text{or} \quad -2 \times 3 = -6 \qquad \text{also} \quad 3 \times -2 = -6$$

But what is $(-2) \times (-3)$?

Remember that $-(-6) = +6$

and $(-2) \times 3 = -6 = -(2 \times 3)$

$\therefore$ $(-2) \times (-3) = -(2 \times -3) = -(-6) = +6$

i.e. the reverse of moving 6 *down* is moving 6 *up*.

The rules for multiplication and division are similar:

Two numbers with like signs give a positive answer.
Two numbers with unlike signs give a negative answer.

EXAMPLE 1

Find the product of $-4 \times -7 \times -3$

$(-4 \times -7) \times -3 = +28 \times -3$
$\qquad\qquad\qquad\qquad = -84$

EXAMPLE 2

Evaluate $(-12 \div 4) \times -6$

$(-12 \div 4) \times -6 = -3 \times -6$
$\qquad = +18$

EXERCISE 1.5

Do *not* use a calculator in the following questions:

1   a -3×6 d $-7 \div -4$ g $8 \div \frac{1}{2} \div -2$ j $\dfrac{20 \times -6}{-4 \times -3}$

    b $15 \div -5$ e $10 \times -3 \div 2$ h $(-3 + -6) \times -4$

    c -7×-4 f $-3 \times -3 \times -3$ i $-7 \div (3 - (-4))$

1.3 Powers and roots

Powers

Patterns of dots like this

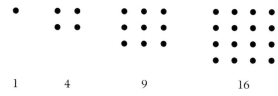

| 1 | 4 | 9 | 16 |

form squares, and the numbers $1, 4, 9, 16, \ldots$ are called the **square numbers**.

The square numbers are also formed by finding the product of an integer with itself:

$$1 \times 1 = 1$$
$$2 \times 2 = 4$$
$$3 \times 3 = 9$$
$$4 \times 4 = 16, \text{ etc.}$$

The sums of odd numbers also produce the square numbers:

$$1 = 1$$
$$1 + 3 = 4$$
$$1 + 3 + 5 = 9$$
$$1 + 3 + 5 + 7 = 16$$
$$1 + 3 + 5 + 7 + 9 = 25, \text{ etc.}$$

If any number is multiplied by itself, the product is called the square of the number. For example:

the square of 1.2 is 1.44
the square of $\sqrt{2}$ is 2.

The square of 3 or 3 squared $= 3 \times 3 = 3^2$, and the 2 is called the **index**.

Similarly:

the cube of $4 = 4$ cubed $= 4 \times 4 \times 4 = 4^3$

and

the fourth power of $2 = 2 \times 2 \times 2 \times 2 = 2^4$

Roots

The square of $4 = 4 \times 4 = 16$.

The number 4, in this example, is called the **square root** of 16, i.e. the number which when squared will give 16.

16 has another square root because $-4 \times -4 = 16$.

Hence the square roots of 16 are $+4$ and -4.

The symbol $\sqrt{\ }$ indicates the positive square root and hence $\sqrt{16} = +4$.

Similarly, the cube root of a number is the number which when multiplied by itself three times equals the given number. For example:

the cube root of 64 or $\sqrt[3]{64} = 4$
(since $4^3 = 4 \times 4 \times 4 = 64$)

EXERCISE 1.6

1   Write down the first 10 square numbers.

2   Write down the first 6 cube numbers.

3   Find the positive square roots of the following numbers:

    a 25 **b** 121 **c** 625 **d** 225 **e** 196

4   **a** Find the cube root of (i) 27 (ii) $^-64$
     (iii) 729

    b Find the fifth root of 32.

    c Find the fourth root of (i) 82 (ii) 625

5   Write 25 as the sum of a sequence of odd numbers.

6   What is the third power of 2?

7   What is 3 cubed?

8   Write down the cube root of:

    a 8 **b** 27 **c** 216 **d** $\frac{1}{8}$ **e** $^-64$

9   What is the fourth root of 81?

10   Evaluate $5^2 \times \sqrt{16}$.

02 Using a Calculator

2.1 Approximations

Most calculators display answers of up to 10 digits.

In most cases, this is too many digits. Therefore, when using a calculator, it is necessary to give an *approximate* answer which contains fewer digits (but which has a sensible degree of accuracy).

There are two methods in common usage:
rounding to a number of **decimal places** (e.g. 2 decimal places or 2 d.p.) and rounding to a number of **significant figures** (e.g. 3 significant figures or 3 s.f.).

The rule for rounding to, for example, 2 decimal places is:

If the digit in the third decimal place is **5 or more**, **round up**, i.e. increase the digit in the second decimal place by 1.

If the digit in the third decimal place is **less than 5**, **round down**, i.e. the digit in the second place remains the same.

EXAMPLE 1

Give the following numbers from a calculator display, to 2 d.p.
7.92341,   25.675231,   0.06666. .,   0.9999999. .

Calculator display	Degree of accuracy required (2 d.p.)	Answer correct to 2 d.p.
7.92341	7.92|341	7.92
25.675231	25.67|5231	25.68
0.066666 .	0.06|6666 .	0.07
0.9999999 . .	0.99|99999 . .	1.00

The rules for significant figures are similar, but you need to take care with zeros.

Zeros at the beginning of a decimal number or at the end of an integer are not counted as significant figures, but must be included in the final result. All other zeros are significant.

For example, 70 631.9 given correct to 3 significant figures is 70 600.

The three significant figures are 7, 0 and 6. The last two zeros are not significant (i.e. do not count as fourth and fifth figures), but are essential so that 7 retains its value of 70 thousand and 6 its value of 6 hundred.

EXAMPLE 2

Give the following numbers, from a calculator display to 3 s.f.
7.92341, 25.675231, 0.066666. ., 24380.., 0.999999. .

Calculator display	Degree of accuracy required (3 s.f.)	Answer correct to 3 s.f.
7.92341	7.92|341	7.92
25.675231	25.6|75231	25.7
0.066666 .	0.0666|66 . .	0.0667
24380.	243|80.	24400
0.9999999. .	0.999|999..	1.00

In the last example, the digit 1 becomes the first significant figure, and the two zeros are the second and third figures.

EXERCISE 2.1

Give each of the following numbers to the accuracy requested in brackets:

1   9.736 (3 s.f.)

2   0.36218 (2 d.p.)

3   147.49 (1 d.p.)

4   28.613 (2 s.f.)

5   0.5252 (2 s.f.)

6   4.1983 (2 d.p.)

7   1245.4 (3 s.f.)

8   0.00425 (3 d.p.)

9   273.6 (2 s.f.)

10   459.97314 (1 d.p.)

2.2   Estimation

The answer displayed on a calculator will be correct for the values you have entered, but a calculator cannot tell you if you have pressed the wrong key or entered your numbers in the wrong order.

Each number you enter into the calculator should be checked for accuracy and the final answer should be checked by comparing it with an *estimated* answer.

EXAMPLE

Estimate the value of 31.41×79.6.

31.41 is approximately 30
79.6 is approximately 80

An estimated value is therefore $30 \times 80 = 2400$.

If the calculator displays, for example, 25002.36, a mistake has been made with the decimal point, and the answer should read 2500.236.

EXERCISE 2.2

1   By rounding all numbers to 1 significant figure, find an estimated value of each calculation:

   a 52.2×67.4 d $607 \div 1.86$ g $\dfrac{520.4 \times 8.065}{99.53}$

   b 6143×0.0381 e $48.2 \div 0.203$

   c 607×1.86 f $3784 \div 412$ h $\dfrac{807}{391.2 \times 0.38}$

2   Find an estimate for each of the following calculations, by choosing an appropiate approximation for each number:

   a $82.3 \div 9.1$ b $0.364 \div 6.29$ c $\dfrac{31.73 \times 6.282}{7.918}$

3   By finding an estimate of the answer, state which of the following calculations are obviously incorrect. (Do not use a calculator.)

   a $8.14 \times 49.6 = 403.74$ f $\dfrac{42.3 \times 3.97}{1635} = 10.27$

   b $23.79 \div 5.57 = 4.27$ g $\sqrt{1640} = 40.5$

   c $324 \div 196 \times 0.5 = 226$ h $\sqrt{650} = 80.6$

   d $3.14 \times 9.46^2 = 882.35$ i $(0.038)^2 = 0.00144$

   e $23.79 \div 0.213 = 11.169$ j $(0.205)^3 = 0.0862$

2.3   Degrees of accuracy

Whenever you solve a numerical problem, you must consider the accuracy required in your answer, especially if you cannot obtain an exact answer. In section 2.1, you saw how to correct an answer to a number of significant figures or decimal places. Section 2.3 shows you how to select an appropriate degree of accuracy.

Sometimes the answer to a numerical problem is an integer, and this gives you an exact answer. Fractions are also exact, but their decimal equivalents are often inexact. For example, suppose you were able to buy sixteen pencils for £3 but wanted to buy only one. Each pencil would cost $£\frac{3}{16}$.

As a decimal this would be 18.75 pence, but obviously you cannot pay 18.75 pence for an individual pencil. The shopkeeper would work in the smallest unit of currency, which is one penny. To make sure that she did not lose money she would round up the price to 19p.

EXAMPLE 1

A netball club is hiring a number of minibuses to go to an away match. Each minibus can carry 15 passengers and 49 members of the club wish to travel. How many minibuses are needed?

The number of minibuses is $\frac{49}{15} = 3.26667$
3.26667 is rounded up to the nearest integer.

The netball club needs 4 minibuses.

Sometimes you will need to **round down** a mathematical answer as the following example shows.

EXAMPLE

Pam changes 9413 French francs into pounds at the rate of 10.9 francs to £1. The bank, not wishing to be generous, decides to round down to the nearest penny the money it will give Pam.

How much does Pam receive?

$$9413 \text{ French francs} = \frac{9413}{10.9} = £863.577\,98$$

∴ Pam receives £863.57

EXERCISE 2.3

1 Twelve pens cost £6.20. What would be charged for one pen?

2 A tin of paint covers $12\,m^2$.
   How many tins are required to cover an area of $32\,m^2$ with two coats of paint?

3 Sara gives each of her 35 friends a chocolate biscuit. Each packet of chocolate biscuits contains 8 biscuits. How many packets of chocolate biscuits must Sara open?

4 An office orders plastic sleeves which cost £2.67 for a box of 100.
   How much should be charged:

   a for ten   b for one?

5 A TV room $5.6\,m^2$ is to be carpeted. The carpet chosen is 3 m wide and is sold in metre lengths. How many metres should be bought?

6 Visitors to a day centre pay £1.70 per week of five days towards the cost of tea and biscuits. How much should someone visiting for one day be charged, if the centre is not to make a loss?

7 Eleven students receive a bill for £120 after an evening out.
   How much should each pay?

8 A booking agency normally sells a block of four tickets for £55.43. It agrees to sell them individually.
   How much should it charge for one ticket if it does not wish to lose money through individual sales?

9 A factory used 26 818 units of electricity at 5.44p per unit.
   What is the cost of the electricity used?

10 In a car factory, pieces of glass fibre 0.9 m are cut from a roll which is 24 m long.
   How many pieces can be cut from one roll?

2.4   The keys of a calculator

To use your calculator most effectively, you must become familiar with the keys and their functions.

The booklet that accompanies your calculator will tell you the order in which the keys are used for calculations.

For example, to find $\sqrt[4]{5}$ the keys required on most calculators are

| 5 | INV | x^y | 4 | = |

Other useful keys are

a the memory keys $\boxed{\text{Min}}$ $\boxed{\text{MR}}$ $\boxed{\text{M+}}$ $\boxed{\text{M}-}$ (can you use them correctly?)

b brackets,   e.g. $\dfrac{36 \times 14}{21 \times 4}$ is $36 \times 14 \div (21 \times 4)$

                    or $36 \times 14 \div 21 \div 4$

                not $36 \times 14 \div 21 \times 4$

c The $\boxed{\text{AC}}$ key clears the calculator (except the memories) before beginning a new calculation.

d The $\boxed{\text{C}}$ or $\boxed{\text{CE}}$ key is used to correct an entry error (i.e. if the wrong key has been pressed).

The key clears the last entry made (either a figure or an operation), provided it is pressed immediately after the error has been made.

The correct entry can then be made and the sequence continued.

For example, $6 + 3$ $\boxed{\text{C}}$ $2 =$ will produce the answer to $6 + 2$.

EXERCISE 2.4

1 $386.9 \div 32.87$ (to 1 d.p.)

2 $2576 \div 0.03741$ (to 3 s.f.)

3 $\sqrt{\dfrac{79.5}{10.9}}$ (to 3 s.f.)

4 0.000356×385.7 (to 3 d.p.)

5 $(7.1 + 4.01) \times 8$ (to 3 s.f.)

6 $\dfrac{1}{0.0345}$ (to 3 s.f.)

7 $(4.63)^2$ (to 3 s.f.)

8 $2.1 + 3.41 \times 7.01$ (to 3 s.f.)

9 $\dfrac{3.8 \times 2.9}{17.1 \times 0.82}$ (to 3 d.p.)

10 $\frac{1}{2}(246.3 + 1092.8 + 376.4 + 49.8)$

11 $\dfrac{3.1 \times 15.2}{7.01 \times 8.11}$ (to 3 d.p.)

12 $\sqrt{\pi \times 38.4}$ (to 2 s.f.)

13 $\sqrt[3]{4 + 3.7 + 28.1}$ (to 1 d.p.)

03 Fractions

3.1 Types of fraction

A fraction is a number which can be written as a ratio, with an integer divided by an integer, e.g. $\frac{7}{9}$ or $-\frac{2}{3}$.

If a shape is divided into a number of equal parts and some of those parts are then shaded, the shaded area can be written as a fraction of the whole.

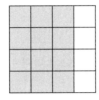

Fraction shaded $= \qquad \frac{3}{8} \qquad\qquad \frac{12}{16} \qquad\qquad \frac{2}{6}$

The **denominator** (lower integer) denotes the number of parts into which the shape was divided.

The **numerator** (upper integer) denotes the number of parts which have been shaded.

Fractions which have a smaller numerator than denominator are called **proper** fractions, e.g. $\frac{1}{2}, \frac{5}{12}, -\frac{28}{60}$.

If the numerator is larger than the denominator, the fraction is an **improper** fraction, e.g. $\frac{12}{5}, -\frac{7}{2}, \frac{72}{18}$.

A **mixed number** is a number composed of an integer and a proper fraction, e.g. $3\frac{1}{2}, -24\frac{3}{4}, 4\frac{5}{6}$.

3.2 Equivalent fractions

We could consider the larger square above to be divided into four columns instead of sixteen squares.

The shaded area is then $\frac{3}{4}$ of the whole. This means that $\frac{12}{16} = \frac{3}{4}$.

$\frac{12}{16}$ and $\frac{3}{4}$ are called **equivalent fractions** because they have the same value.

EXAMPLE 1

Complete $\dfrac{6}{7} = \dfrac{?}{28}$ to give equivalent fractions.

The 7 in the denominator must be increased four times to give 28.

The numerator must be treated similarly and be increased four times:

$$\frac{6}{7} = \frac{6 \times 4}{7 \times 4} = \frac{24}{28}$$

EXAMPLE 2

Reduce $\dfrac{32}{72}$ to its lowest terms.

$$\frac{32}{72} = \frac{32 \div 8}{72 \div 8} = \frac{4}{9}$$

(4 and 9 have no common factor therefore the fraction is in its lowest terms.)

EXAMPLE 3

Convert $13\frac{2}{5}$ to an improper fraction.

$$13\frac{2}{5} = 13 + \frac{2}{5}$$
$$= (13 \times \frac{5}{5}) + \frac{2}{5}$$
$$= \frac{65}{5} + \frac{2}{5}$$
$$= \frac{65 + 2}{5}$$
$$= \frac{67}{5}$$

EXAMPLE 4

Convert $\dfrac{141}{22}$ to a mixed number.

$$22)\overline{141}$$
$$\qquad\qquad 6 \text{ remainder } 9$$

$$\frac{141}{22} = 6\frac{9}{22}$$

EXERCISE 3.1

1   Complete each of the following to give equivalent fractions:

   a $\dfrac{4}{5} = \dfrac{?}{10}$ **c** $\dfrac{3}{4} = \dfrac{?}{24}$ **e** $\dfrac{21}{30} = \dfrac{?}{10}$

   b $\dfrac{2}{3} = \dfrac{?}{12}$ **d** $\dfrac{1}{4} = \dfrac{?}{36}$ **f** $\dfrac{16}{18} = \dfrac{8}{?}$

2   Reduce the following fractions to their lowest terms:

   a $\dfrac{10}{15}$ **b** $\dfrac{18}{24}$ **c** $\dfrac{22}{99}$ **d** $\dfrac{21}{33}$

   e $\dfrac{75}{100}$ **f** $\dfrac{40}{60}$ **g** $\dfrac{30}{65}$ **h** $\dfrac{64}{96}$

3   A particular shade of green paint is made by mixing 60 g of blue powder paint with 30 g of yellow powder paint.
What fraction of the mixture is blue paint?

4   A secretary's working day is 8 hours long. On one particular day, $4\frac{1}{2}$ hours were spent typing and $1\frac{1}{2}$ hours on the telephone.
What fraction of the working day was spent:

   a typing   **b** telephoning?

5   A dental chart shows that a patient with a full set of 32 teeth has 6 fillings.
What fraction of the teeth are filled?

6   In a group of 240 tourists travelling on a charter flight, 180 requested seats in the non-smoking section.
What fraction of the passengers were in the smoking section?

7   A printing firm produces 350 books in a day. 200 of these are to be exported.
What fraction of the day's production is exported?

8   Convert the following mixed numbers to improper fractions:

   a $2\dfrac{2}{9}$ **b** $5\dfrac{1}{6}$ **c** $4\dfrac{11}{12}$ **d** $11\dfrac{1}{10}$

   e $8\dfrac{3}{4}$ **f** $12\dfrac{2}{3}$ **g** $7\dfrac{3}{20}$ **h** $20\dfrac{3}{7}$

9   Convert the following improper fractions to mixed numbers, in their lowest terms:

   a $\dfrac{9}{7}$ **b** $\dfrac{36}{5}$ **c** $\dfrac{30}{9}$ **d** $\dfrac{61}{8}$

   e $\dfrac{153}{11}$ **f** $\dfrac{127}{12}$ **g** $\dfrac{132}{15}$ **h** $\dfrac{94}{4}$

3.3   Operations involving fractions

Addition and subtraction

Only fractions *of the same type* can be added or subtracted, i.e. they must have the same denominator.

The method for addition or subtraction is:

(i)   Find the smallest number which is a multiple of all the denominators.

(ii) Change each fraction to an equivalent fraction with the new denominator.

(iii) Add and/or subtract the fractions.

(iv) If the answer is an improper fraction, convert to a mixed number.

(v)   Give the answer in its lowest terms.

EXAMPLE 1

Evaluate $\dfrac{5}{8} - \dfrac{3}{4} + \dfrac{1}{5}$

40 is the smallest number that is a multiple of all the denominators.

The sum in equivalent fractions is $\dfrac{5 \times 5}{8 \times 5} - \dfrac{3 \times 10}{4 \times 10} + \dfrac{1 \times 8}{5 \times 8} = \dfrac{25}{40} - \dfrac{30}{40} + \dfrac{8}{40}$

$= \dfrac{3}{40}$ which is a fraction in its lowest terms.

EXERCISE 3.2

Evaluate the following:

1 $\dfrac{2}{5} + \dfrac{3}{4}$ 2 $\dfrac{7}{8} - \dfrac{5}{6}$ 3 $\dfrac{5}{12} + \dfrac{1}{4}$ 4 $1\dfrac{1}{2} - \dfrac{3}{4}$

5   A clear glaze for pottery is made by mixing feldspar, flint, whiting and china clay.
One half of the mix is feldspar and one fifth is china clay. Flint and whiting are mixed in equal amounts.
What fraction of the mix is flint?

6   a Photographic prints $4\dfrac{1}{2}$ in wide by $3\dfrac{1}{2}$ in are to be mounted in an album which has pages $9\dfrac{3}{4}$ in wide by $13\dfrac{1}{4}$ in.
How many prints can be mounted on one page?

   b The photographs are to be equally spaced on the page. What width are the margins:

   a across the page   b down the page?

7   Three quarters of an office's stationary budget is spent on paper, one sixth on envelopes, and the remainder on miscellaneous items.
What fraction is spent on miscellaneous items?

8   To encourage customers to pay their bills, a firm gives a discount of $\dfrac{1}{20}$ of the bill if it is paid on time and a further discount of $\dfrac{1}{12}$ of the bill for early payment.
What fraction of the bill is deducted for early payment?

9   A dietician, advising clients on suitable diets, found that $\dfrac{3}{4}$ of her clients were overweight, $\dfrac{1}{6}$ suffered from arthritis and the remainder from coeliac disease.

   a What fraction were coeliac sufferers?

   b If her clients numbered 36 at the time, how many needed a gluten-free diet?

10 A cold remedy is sold as a powder in sachets. $\frac{4}{5}$ of each powder is aspirin and $\frac{2}{25}$ is ascorbic acid. The remainder is caffeine.
   a What fraction is caffeine?
   b How much of each ingredient does 500 mg of powder contain?

11 The proprietor of a B & B establishment mixes his own breakfast cereal. This consists of $2\frac{1}{2}$ cups of oats, $\frac{1}{4}$ of a cup of wheat germ, and $\frac{2}{3}$ of a cup of raisins and nuts.
   What is the total number of cups in this mixture?

12 During a three-day holiday break, $1\frac{1}{2}$ in of rain fell on the first day, $1\frac{3}{4}$ in on the second day and $3\frac{3}{4}$ in on the third day.
   How much rain fell altogether?

13 An axle of diameter $2\frac{3}{4}$ inches is fitted into the centre of the hub of a wheel which has a diameter of $3\frac{1}{8}$ inches.
   How much clearance is there between the axle and the inside of the hub on each side?

14 In a self-assembly unit, a wood top, $\frac{5}{8}$ in thick, is screwed to a metal frame $1\frac{1}{4}$ in thick.
   What is the maximum length of screw that can be used?

3.4   The conversion between fractions and decimal fractions

The fraction $\frac{7}{8}$ may be stated as $7 \div 8$.

Using a calculator, $7 \div 8 = 0.875$, and this is the **decimal fraction** which is equivalent to $\frac{7}{8}$.

EXAMPLE 1

Convert $3\frac{5}{6}$ to a decimal.

The integer part of the mixed number remains the same. Only the fractional part needs to be converted.

On the calculator $5 \div 6 = 0.833\,3333$ which is a recurring decimal.

$$3\frac{5}{6} = 3.8\dot{3} \text{ or } 3.83 \text{ (to 3 s.f.)}$$

All fractions convert to either a terminating or a recurring decimal. The dot above the 3 indicates that the 3 is a recurring decimal.

EXAMPLE 2

Convert 0.35 to a fraction in its lowest terms.

$$0.35 = \frac{35}{100} = \frac{7}{20} \text{ (dividing numerator and denominator by 5)}$$

EXERCISE 3.3

1 Write down the shaded area as (i) a fraction, (ii) a decimal fraction of the whole area.

a   **b**   **c**

2 Convert the following fractions to decimals:

a $\dfrac{1}{10}$ **b** $\dfrac{1}{2}$ **c** $\dfrac{3}{4}$ **d** $1\dfrac{9}{20}$

e $4\dfrac{21}{25}$ **f** $2\dfrac{5}{6}$ **g** $7\dfrac{4}{9}$ **h** $3\dfrac{1}{7}$

3 Convert the following decimals to fractions:
   a 0.5   **f** 2.8
   b 0.25   **g** 3.6
   c 1.$\dot{6}$ **h** 2.15
   d 1.$\dot{3}$ **i** 0.125
   e 1.3   **j** 0.375

4 Investigate the relationships between the fractional and decimal forms of:

   a Halves, quarters and eighths

   b thirds, sixths and ninths.

5 Describe a quick method of converting to decimals:

   a tenths   **b** fifths   **c** hundredths

   d twentieths   **e** twenty-fifths.

In questions 6–11, where appropriate, write the answer:

 (i) as a mixed number in its lowest terms

(ii) as a decimal correct to 2 d.p.

6 A designer draws a sketch for a new design of car. The length of the car on the sketch is $4\frac{1}{2}$ inches. The actual length of the car is 153 inches. How many times larger is the actual car than the sketch?

7 **a** A chemist sells toothpaste in two sizes: 75 g and 125 g.
   How many times larger is the 125 g tube than the 75 g tube of toothpaste?

   b If the cost of the 75 g tube of paste is 48p what should the equivalent price of the 125 g tube be?

8 Traditionally, 1 quire of paper = 24 sheets
                    1 ream of paper = 20 quires
Nowadays, however, a ream is generally 500 sheets.

   a How many extra sheets are there in a ream?

   b How many quires are there in a ream?

9 On a hospital ward, 40 minutes is spent every day checking patients' temperatures and blood pressures.

   How many hours are spent on this activity in a 7-day week?

10 A tour bus driver has to travel 112 miles on the first leg of the journey. The driver expects to travel at an average speed of 35 mph.

   What is his estimate of the time for this part of the journey?

11 A piece of machinery has two interlocking cogs.
   Cog A has 35 teeth.   Cog B has 20 teeth.

   How many turns does Cog B make for each turn of Cog A?

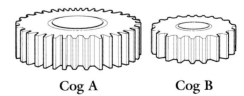

   Cog A   **Cog B**

04 Percentages

'Inflation now stands at 2.3%.' 'Ford have given their workforce a 3.5% rise.'
'Unemployment in Winchester is less than 2%.'

Percentages are a part of our everyday lives. They are often quoted in the media, particularly in connection with money matters.

Percentages often help us make comparisons between numbers, but we must know exactly what a 'percentage' is.

4.1 Percentages

A percentage is a fraction with a particular number divided by 100:

$$20\% \text{ means } \frac{20}{100}$$

A decrease of 20% would be a decrease of $\frac{20}{100}$, which is the same thing as a decrease of $\frac{1}{5}$, a fifth.

$$\frac{1}{4} = \frac{1}{4} \times \frac{25}{25} = \frac{25}{100} \text{ which is } 25\%$$

To convert a fraction to a percentage multiply by 100.
To convert a percentage to a fraction, divide by 100.

EXAMPLE 1

Convert   **a** $\frac{3}{5}$, **b** $\frac{9}{11}$ to percentages.

a $\frac{3}{5}$ as a percentage $= \frac{3}{5} \times 100 = 60\%$

b $\frac{9}{11}$ as a percentage $= \frac{9}{11} \times 100 = 81.82\%$ (to 2 d.p.)

EXAMPLE 2

Convert 65% to a fraction in its lowest terms.

$$65\% = \frac{65}{100} = \frac{13}{20}$$

---

1 Convert the following fractions to percentages:

a $\frac{1}{5}$ b $\frac{1}{8}$ c $\frac{7}{10}$

d $\frac{13}{20}$ e $\frac{2}{3}$ f $\frac{9}{25}$

g $1\frac{3}{4}$ h $2\frac{1}{2}$

2 Convert the following percentages to fractions:

a 60%   b 25%   c 10%

d 85%   e 15%   f 130%

g $37\frac{1}{2}\%$ h $33\frac{1}{3}\%$

3 Copy and complete the following table to give each quantity in its fractional, decimal and percentage form.

	Fraction	Decimal	Percentage
a	$\frac{3}{4}$		
b		0.5	
c	$\frac{1}{8}$		
d			$33\frac{1}{3}$
e		0.375	
f	$\frac{7}{10}$		
g			35
h		0.6	
i	$\frac{3}{5}$		
j			62.5

4.2 Finding a percentage of an amount

EXAMPLE 2

Julian reads in the newspaper that the average pocket money for 12-year-olds has increased nationally by 14% in the last year.
Julian's 12-year-old daughter Gilly has been given £1.60 per week for the last two years. How much more per week should he provide for a 14% increase?

$$14\% = \frac{14}{100}$$

$$14\% \text{ of } £1.60 = \frac{14}{100} \times £1.60$$

$$= 0.14 \times £1.60 = 22.4\text{p}$$

Increase in pocket money = 22.4p = 22p (to the nearest 1p)
Note. See Section 2.1 for a further explanation of approximations.

EXERCISE 4.2

1   Calculate the following percentages to the nearest 1p.

   a   10% of £13.75   e   123% of £4.20
   b   50% of £637.24   f   121% of £69.80
   c   7% of £316   g   10.80% of £900
   d   15% of £92.72   h   34.3% of £128.50

4.3   Increasing an amount by a given percentage

EXAMPLE

TRAIN FARES TO RISE BY 6%

What does this mean in cash terms to the 20 000 long-distance commuters who travel to London every day?

If an InterCity season ticket costs £2632 now, how much will it cost after the rise?

Method 1
To find the new cost of a ticket we can find 6% of £2632 and then add this to the original:

$$6\% \text{ of } £2632 = \frac{6}{100} \times £2632$$

Increase in fare = £157.92

New cost of fare = £2632 + £157.92

$$= £2789.92$$

Method 2
Consider the original amount of £2632 as 100%. Increasing it by 6% is the same as finding 106% of the original cost.

$$106\% = \frac{106}{100} = 1.06$$

Therefore the quickest way of increasing the original fare by 6% is to multiply it by 1.06.

$$106\% \text{ of } £2632 = 1.06 \times £2632$$

New cost of fare = £2789.92

EXERCISE 4.3

Give all answers to the nearest 1p.

1   Increase the following rail fares by 10%.

   a   £9.20   d   £9.81
   b   £3.70   e   £9.13
   c   £5.00

2   Increase the given amount by the required percentage.

   a   £72.12 by 50%   d   £220 by $6\frac{1}{4}\%$
   b   95p by 10%   e   £124.80 by 25%
   c   £360 by 120%   f   £19.99 by $8\frac{1}{2}\%$

4.4 Decreasing an amount by a given percentage

EXAMPLE

The marked price of this sweater is £24.90.
What is its sale price?

Method 1

The reduction = 20% of £24.90

$$= \frac{20}{100} \times £24.90$$

$$= £4.98$$

The sale price = £24.90 − £4.98

$$= £19.92$$

Method 2

£24.90 is the equivalent of 100% and so decreasing the price by 20% is equivalent to finding 80% of the original price.

$$\frac{80}{100} \times £24.90 = 0.80 \times £24.90$$

Sale price = £19.92

EXERCISE 4.4

Give all answers to the nearest 1p.

1 Reduce the following marked prices by 20% to find the sale prices:

a £30.00   d 45p

b £10.50   e £12.99

c £17.60

2 Decrease the given amount by the required percentage:

a £54.10 by 8%   d £27.15 by $12\frac{1}{2}$%

b 84p by 30%   e £99.05 by 40%

c £128 by 60%   f £1.62 by 33%

4.5 Expressing one quantity as a percentage of another

In a survey of insurance companies it was found that the most common type of car accident was one car running into the back of another.

Out of 35 000 claims, 6280 were for this type of accident.

Information of this type is usually quoted as a percentage.

EXAMPLE 1

Find 6280 as a percentage of 35 000.

First express 6280 as a fraction of 35 000:

$$\frac{6280}{35\,000}$$

Then multiply this fraction by 100 to express it as a percentage:

6280 as a percentage of 35 000

$$= \frac{6280}{35\,000} \times 100\%$$

$$= 17.9\% \ (3 \ \text{sf})$$

This means that almost 18% of car accidents are caused by cars running into the backs of other vehicles.

Note. See Unit 2 for a further explanation of significant figures.

EXAMPLE 2

A shop buys wallpaper from a wholesaler at £6 per roll and sells it to customers at £8.20 per roll.

What is the percentage increase in price?

The **increase** in price is £8.20 − £6.00 = £2.20

This is $\dfrac{£2.20}{£6.00}$ as a fraction of the **original** price.

$$\text{Percentage increase} = \frac{£2.20}{£6.00} \times 100\%$$

$$= 36.7\%$$

EXAMPLE 3

By what percentage has the marked price of £4.70 been decreased to give a sale price of £3.80?

The **decrease** in price is expressed as a fraction of the **original** price and then multiplied by 100.

$$\text{Decrease in price} = 90\text{p}$$

$$\text{Percentage decrease} = \frac{90\text{p}}{£4.70} \times 100\%$$

$$= \frac{£0.90}{£4.70} \times 100\% \ (\text{both quantities must be in the same units})$$

$$= 19.1\%$$

EXERCISE 4.5

Give all answers correct to 3 significant figures.

1 Express the first quantity as a percentage of the second:

    a 20, 25   **d** 54, 108

    b 3, 87   **e** 60p, £1.10

    c 140, 80   **f** £16.25, £12.50

2 Find the percentage by which the first amount is increased or decreased to give the second amount:

    a £65, £80   **d** £499, £399

    b £250, £300   **e** £1.23, 67p

    c 20p, 95p   **f** £24.50, £138.20

EXERCISE 4.6

1   A dress made by a Paris Fashion House has a recommended retail price of £870. A London shop advertises it at £710.
By what percentage has the shop reduced the price?
Give your answer to the nearest integer.

2   Linda and Mario are discussing the students at their Art College.

a   Linda says that 60% of the students are female and Mario says that there are 52 more females than males.
How many students are there at the college?

b   Linda says that $\frac{2}{5}$ of the students attend on Wednesday only in the morning and Mario knows that 30% of the students attend all day on Wednesday.
How many students do not go into college on Wednesday?

3   A watch has a MRRP (maker's recommended retail price) of £32.99, but a jeweller's shop advertises it for £28.99.
By what percentage has the shop reduced the price (to the nearest whole number)?

4   In a town of 25 000 inhabitants, 80% are over 18 years of age.

a   How many of the inhabitants are over 18?

Of these, 37% usually shop in the town's supermarket.

b   How many shop in the supermarket?

7240 people over 18 living in the town use hypermarkets regularly.

c   What percentage of people over 18 shop in hypermarkets?

5   A survey on teenage smoking found that 70% of girls of secondary school age tried smoking and that 36% of those who tried it became addicted.

In a secondary school with 580 female students:

a   how many girls would you expect to find had tried smoking?

b   how many girls would you expect to find had become addicted to smoking?

6   The 36 residents in a rest-home each pay £231 per week. 28% of the total income of the home is spent on nursing care. The rest-home employs its nurses for a total of 588 hours per week.
What is the hourly rate of pay for each nurse?

7   Restaurants often add a service charge of 12% to your bill. A meal for two costs £28.60.
How much service charge will be added?

8   'Sunny Tours' offers a 5% discount on all holidays booked before 31 December the previous year.

How much will a family of 2 adults and 2 children aged 11 and 15 pay for a holiday whose advertised price is £380 each with a 30% reduction for children under 14 years of age?

9   In the survey of 3500 accidents at work, 17.6% happened on a Friday.
How many of the accidents occurred on a Friday?

10   Mr Robinson invested £32 000, together with money raised from a loan from his bank, in a manufacturing enterprise. He spent 65% on buildings, with an additional 28% on equipment. How much did he have left to spend on materials?

05 Statistical Terms

In 1834, the Royal Statistical Society was founded, and defined statistics as 'using figures and tabular exhibitions to illustrate the conditions and prospects of society'. Statistics is now used to deal with the collection, classification, tabulation and analysis of information and opinions.

Data. Data is the information which has been collected or researched. The word 'data' is a plural and the singular is 'datum' (a single piece of information).

Variables. Information is collected about *variables* such as weights, numbers of clients, types of disease.

A *variable* is something which can change from one item to the next. It can be either **quantitative** (i.e. numerical like weight or number of clients) or *qualitative* (i.e. an attribute like car colour or type of disease).

There are two types of quantitative variables:

  (i) *Continuous.* A continuous variable is a variable which could take all possible values within a given range, e.g. the height of a tree.

  (ii) *Discrete.* A discrete variable is a variable which increases in steps (often whole numbers), e.g. the number of rooms in a building.

A discrete variable does not have to consist only of whole numbers. For example, the size of shoes is also a discrete variable, and the sizes go up in steps of a half (5, $5\frac{1}{2}$, 6, $6\frac{1}{2}$, etc.).

The number of steps climbed is a *discrete* variable.

The distance travelled on the escalator is a *continuous* variable.

Observation. An observation is the value taken by a variable. For example, an age of 17 years is an observation when the variable is age.

Population. The term *'population'* means everything (or everybody) in the category you are considering. For example, if you were making a study of cathedrals, the population could be all the cathedrals in Britain. If you were investigating what attracts people to certain types of holiday, the population would be all holiday-makers.

EXERCISE 5.1

For each population below, state whether the variable given is qualitative, discrete or continuous:

1   the numbers of employees in a county's factories

2   the weights of new born babies in Britain

3   the age at death in 1994 of a town's inhabitants

4   the lengths of bolts coming off a factory production line

5   the colour preferences of customers in a city's clothes shops

6   the acceleration rates of new models of motorbike in a given year

7   the time taken to complete a job by each employee of a firm

8   the brands of toothpaste sold by chemists

9   the number of passengers on flights to the Continent during one summer

10   the newspapers on sale at station kiosks

11   the number of cars parked each morning in a firm's car park over a period of time

12   the type of holiday accommodation available in a resort.

06 Sampling, Surveys, Questionnaires

One of the problems with statistical surveys involving people is that, whatever your opinion, there are likely to be many other people with the same opinion. If you ask only these people, your opinion will be seen to be that of the whole population. If you ask only people with the opposite opinion, you will be seen to be in a minority. Therefore, you must ask a variety of people, so that you have a true picture of the population.

Remember, however, that in statistics, the term **population** does not necessarily refer to people. If you wish to survey the ages of cars on the road, your population might be all the cars in Britain.

6.1 Surveys

When you record any information – for example, about other people's opinions or numbers of surviving African elephants or types of road accidents – you are carrying out a **survey**. The survey results may be obtained by asking questions, by observation or by research.

To obtain completely accurate information, you would have to ask *everybody* (in your town or country or whatever), and receive answers from everybody, or observe *all* the elephants in Africa.

6.2 Censuses

When information is gathered about all the members of a population, the survey is called a **census**.

A national census is carried out every ten years, in years ending with 1 (e.g. 1991, 2001). Every adult in Britain is asked a large number of questions on mainly factual matters, for example the number of rooms in their house, their age, and the number of cars they possess.

A national census is a very large undertaking, and the results, though accurate, take a substantial length of time to be produced. Apart from the vast number of people to be asked, and the placing of their answers in computers, it is very difficult to ensure that every adult has in fact replied. It costs the country a great deal of money to complete a national census.

6.3 Samples

It is usually impossible for firms, newspapers, biologists, medical researchers, etc., to obtain information about the whole population, because the survey:

- may be expensive
- may take a long time
- may involve testing to destruction – e.g. if you wish to find out how long batteries last, you test them until they run out
- may be impossible to carry out for every member of the population – e.g. a survey to find the weights of trout in Scottish rivers.

A small part of the population is chosen for the survey and this is called a **sample**.

The statistician then assumes that the results for the sample are representative of the population as a whole. The larger the number of people asked, the more likely their response is to be a valid result for the whole population.

Clearly it is vital that for the survey to be accurate the sample you choose must be representative of the whole population.

To achieve this, every member of the population must have an equal chance of being chosen.

6.4   Sampling methods

Random sampling

A random sample is one in which every member has an equal chance of being selected.

Campaign groups for or against a particular issue (such as the possible siting of a new supermarket near a park) can often obtain a large majority for their point of view simply by selecting which passers-by to question (perhaps the people living near the park who will be worried about the possibility of noise). By careful selection, majorities as high as 70% can easily be obtained both for and against the same issue! (Some people may well want a supermarket behind their back garden.)

The simplest way to obtain a random sample is to give every member a number, and to select numbers from tickets in a box (as in a raffle), or (if there are too many for this method) to select numbers by computer. Random numbers can also be obtained by using the RAN button on some calculators.

It is common to use the electoral roll of a suitably sized area (on which every adult is listed) to obtain a numbered list from which to select a sample.

Periodic sampling

With periodic or systematic sampling, a regular pattern is used to pick the sample, for example, every hundredth firework on a production line. This can give a unrepresentative sample if there is a pattern to the list which is echoed by the sample.

Stratified random sampling

A stratified sample is more accurate than a random sample, and is used in opinion polls, when 1 or 2% accuracy is important. A stratified sample (or **strata sample**) is one in which the population is divided into categories. The sample should then be constructed to have the same categories in the same proportions.

Random sampling is then used to select the required numbers in each category.

For example, if you wished to find out about the earnings of students in a sixth-form college, it would be sensible to have both lower sixth and upper sixth students represented. You may also wish to make sure that one-year students, males and females, are fairly represented. Suppose there are 1000 students in college, of whom 220 are lower sixth one-year students, 420 are lower sixth two-year students and 360 are upper sixth students.

A sample of 50 would contain the following numbers:

LVI one-year students $= \dfrac{220}{1000} \times 50 = 11$

LVI two-year students $= \dfrac{420}{1000} \times 50 = 21$

UVI students $\qquad = \dfrac{360}{1000} \times 50 = 18$

The eleven LVI one-year students would be randomly chosen from the 220 students in college. The other two strata would be chosen in the same way.

Quota sampling

For a quota sample, a manufacturer may determine the proportions of each group to interview.

For example, if a manufacturer wishes to launch a new chocolate bar on the market, it may be more important to canvass the opinions of children and those who do the shopping than any other sector of the market.

A market researcher paid to survey a sample of 100 people could be instructed to ask, say, 20 people under the age of 18, 30 in the age range 19–40 who do the family shopping, 10 in the same age range who don't, 30 in the age range over 40 who do the family shopping, and 10 in this age range who don't. The researcher will probably use convenience sampling (see below) to choose who to ask, but once one of the quotas is filled, no more people in that category may be asked. The researcher will continue to ask people in the other categories until the sample of 100 has been surveyed.

This is a common method used for market research, but inexperienced (or lazy!) researchers may choose an unrepresentative sample.

Convenience sampling

The most convenient sample is chosen, which, for a sample of size fifty, usually means the first fifty people you meet. There is obviously no guarantee that this sample will be representative. In fact it is highly likely that it won't be.

6.5   Bias

The results of a survey are biased if the sample is not representative of the whole population.

Bias can be introduced if:

- the sample is unrepresentative. Even when using random sampling an unusual sample may be chosen, and this is just bad luck.
- an incorrect sampling method is used. Sampling methods, other than random, or stratified random sampling, are very likely to produce biased samples.

If you wanted to know people's views on drinking, a survey held outside a public house at closing time would clearly produce a different response from one held outside the office of the 'Teetotallers' League'! Neither would be representative of the complete population. Both of these samples would be biased.

- the questions asked in the survey are not clear or are leading questions (see section 6.6).

EXERCISE 6.1

In questions 1 to 8:

a identify the population

b criticise the method of obtaining the sample

c recommend an alternative way of obtaining a sample.

1   A journalist at a local newspaper wants to canvass popular opinion about plans for a new shopping centre in town. He goes into the High Street, and asks people, until he has asked 50.

2   Stephanie wishes to find out the earnings of college students. She goes into a college common room, and asks 40 girls.

3   The police wish to ascertain how many cars have a valid tax disc. One day, they set up a survey point on a road out of a town, between 5pm and 6pm. They stop a car and check its tax disc. As soon as it has left, they stop the next car.

4   A geography student needs to collect five soil samples from his garden for a project. He stands in the middle, and throws a coin in the air. Where it lands, he takes a sample.

5   For a survey into the smoking habits of teenagers, Carol went to a tobacconist's near a school at 3.30pm, which was when the school day ended. She asked everyone entering the shop how much they spent on cigarettes in a week.

6   To find out how many homes in a telephone area have central heating, a salesgirl telephones 100 people, picked at random from a telephone directory.

7   To find out the make of car that people in an area of town use, Peter went out after lunch and knocked on doors until he had one hundred responses. He was pleased with his efficiency, as he had finished by 4pm.

8   To investigate what influenced people in their decision on mode of transport to work, John went to the station just before the 8.15 train departed, and asked as many people as he could.

6.6 Questionnaires

If your questions are written down and given to people to complete, the list of questions is called a **questionnaire**. The questions you ask must be chosen with care. They must:

(i)   **not give offence.** Some people do not wish to give their precise age, or social class, so you *either*: (*a*) find an alternative questions, (e.g. 'Which of these age ranges applies to you?'), or (*b*) fill in the information by using your own judgement.

(ii) **not be leading.** 'What do you think of the superb new facilities at . . .' will *lead* most people to agree they are better than the old facilities. People do not usually want to contradict the questioner. However, the point of the survey should not be to obtain agreement with your view, but to obtain other people's opinions.

(iii) **be able to be answered quickly.** The person answering the questions will often have only a small amount of time to spare and will not want you to write long sentences on their point of view. To obtain information easily from the survey it is helpful to have Yes/No answers or 'boxes' for the answers which are ticked. Here is an example:

How many different television sets does your household possess?

0	1	2	3	4	5	More
☐	☐	☐	☐	☐	☐	☐

A questionnaire must also be easy for anyone to understand. The questions themselves must also be designed carefully. The question 'How much do you watch TV?' could result in the following types of response:

'A lot', 'Not much'
'Every night', 'twice a week'
'For two hours a night'
'Whenever there's sport, a film, . . .'

A better question is 'How many hours do you spend watching TV?', but this may encourage wild guesses because of poor memory.

An even better question to ask is 'How many hours did you watch TV *last* night?'. You can then offer a range of possible answers such as:

'Not at all'
'Up to $\frac{1}{2}$ hour'
'$\frac{1}{2}$ to 1 hour'
'1 to 2 hours'

and so on.

If you suspect different times are spent on different days, it is up to you, as a statistician, to ask a few people each day over a period of a week.

All surveys are open to error. The larger the sample, the more accurate the result.

6.7   Pilot surveys

It is common for companies to carry out an initial survey on a small area of the country in order to identify potential problems with the questions and to identify typical responses. This limits the errors in expensive large-scale surveys.

EXERCISE 6.2

Criticise the questions asked in this exercise and suggest questions which should be asked to find the information required.

1   What do you think of the improved checkout facilities?

2   Do you agree that BBC2 programmes are the best on TV?

3   What is your date of birth?

4   Sheepskin coats are made from sheep. Do you wear a sheepskin coat?

5   Dolphins are wild animals. Do you enjoy watching dolphins perform?

6   Sunbathing causes skin cancer. Do you sunbathe?

7   Vitamin D is obtained from sunlight. Do you sunbathe?

8   Is the new decor a major improvement on the old?

9   Would you rather use your local shops than a major supermarket miles away?

6.8   Hypothesis testing

A statistical survey should have a purpose. It may be used to find out people's opinions (e.g. an opinion poll) or to discover what the population requires of a new product (consumer research), but often it is used to test a theory. A statement is made about a population, or populations, which can be tested statistically. This statement is called a **hypothesis**.

For example:

1 'Women live longer than men.'

2 'You can't tell margarine from butter.'

3 'A new leisure centre would benefit the town.'

4 'The most popular colour of car is red.'

are four statements which are **hypotheses** (plural of 'hypothesis'). Each of these hypotheses can be investigated statistically.

To test the truth of the hypothesis someone must devise an appropriate method to collect data. This must then be analysed before a conclusion can be made, based on the results of the analysis.

The hypothesis can be tested by carrying out a survey or experiment, by observation, or by using published data (which is the result of someone else's survey).

For the four examples above:

1 To test the hypothesis that women live longer than men, government statistics published over several years could be used. Averages and measures of spread could be calculated (see Unit 9) and compared.

2 An experiment could test whether or not it was possible to tell the difference in taste between margarine and butter, perhaps by blindfolding people and seeing whether they can identify which is which.

3 A method for deciding if a new leisure centre is needed could be to devise a suitable questionnaire and survey a sample of the town population.
Among other things, the questionnaire would need to find out what facilities people required for sports, how their needs were being met at the present time, and whether they would consider using a new local leisure centre.

4 One method of testing the hypothesis that 'the most popular colour of car is red' would be to carry out a survey of cars on a busy stretch of road and record results taken at different times on different days on a survey sheet.

EXERCISE 6.3

1 State an appropriate method which could be used to test the following hypotheses:

   a If it rains on St Swithun's day, it will rain for the next forty days and forty nights.

   b Consumers prefer . . .
(Choose any product or set of products which interests you).

2 Devise an experiment to test the hypothesis: 'Students studying Leisure and Tourism have quicker reactions than those studying Art and Design.'

3 An artist wants to find out the preferences of potential customers in their choice of paintings. How would he carry out a survey to discover these preferences?

4   A potter in Cornwall considers whether to produce coffee mugs or cups and saucers for his new hand-made range.
    How would he carry out a survey to test opinion on which holiday-makers would prefer to buy?

5   Design a questionnaire to test the hypothesis that most burglar alarms are sold to victims of a recent burglary.

6   An estate agent wants to carry out a survey to discover what incentives would make house sellers use his agency in preference to others. How would he do this?

7   One village has a high incidence of childhood leukaemia.
    How would you test the hypothesis that this is due to natural causes?

8   A student nurse decides to investigate differences in the occurrence of breast cancer in Europe. The rate in Southern France is very low. Design a questionnaire to test the hypothesis that this difference is due to diet rather than lifestyle.

9   A tour operator wonders whether it would be profitable to arrange flights between Exeter, the local airport, and Malaga, Spain.
    Construct three questions for a questionnaire designed to discover whether this would be profitable.

10 A doctor analyses sporting injuries in two neighbouring villages, Hartington and Easeham. Hartington shows a greater number than Easeham. The doctor believes that this is because the residents of Hartington are younger than those in Easeham.
    How would you design a questionnaire to test this hypothesis?

11 A company considers producing a cabriolet (open-top) version of a small car.
    Design a questionnaire to survey opinion and discover whether this would be a sensible decision.

12 A manufacturer makes fittings for front doors. The fittings are made in brass and chrome and the manufacturer is wondering whether to introduce a range in black matt finish.
    How would the company carry out a survey to find out whether the new range would be successful?

07 Classification and Tabulation of Data

7.1 Tabulation

The purpose of tabulation is to arrange information, after collection and classification, into a compact space so that it can be read easily and quickly. It then may be represented pictorially to enable relevant facts to be seen readily, as explained in the next chapter.

Tabulation consists of entering the data found in columns or rows.

EXAMPLE

The numbers of pensioners living in certain villages were:

Village	Number of pensioners
Ashurst	31
Botleigh	17
Crow	28
Downton	24
Eaglecliffe	19
Fillingdales	33
Total	152

It is important that the tables produced are neat, all rows and columns are clearly identified, and that units (where appropriate) are given.

7.2 Classification of data

Assuming that additional data had been collected, more detailed information could be given by subdividing the rows and/or columns.

EXAMPLE

Using the data from the Example above and subdividing the columns into male and female gives more information about the pensioners:

Village	Number of pensioners	
	Male	Female
Ashurst	12	19
Botleigh	5	12
Crow	10	18
Downton	11	13
Eaglecliffe	9	10
Fillingdales	15	18
Total	62	90

Note: It is essential that all the relevant information is collected during the survey.

It is not possible, for example, to determine a person's sex after the survey has been completed.

7.3   Tally charts

It is common to record the data by means of a **tally chart**.

Suppose a survey was being carried out to determine the popularity of the various activities offered at a local leisure centre. First a list would be drawn up of possible activities: swimming, badminton, fitness training, etc. Then each person entering the leisure centre would be asked which activity they were paying for and a tally mark (*I*) would be recorded against the chosen activity.

To enable the results to be totalled quickly, it is usual to tally in groups of five, the fifth stroke being drawn diagonally across the previous four: JHT .

A section of the results for this survey could look like this:

Activity	Tally	Total
Archery	JHT JHT I	11
Badminton	JHT JHT JHT JHT III	23
Bowls	JHT III	8
Fitness room	JHT JHT JHT IIII	19
Judo	JHT JHT JHT JHT	20

7.4   Frequency tables

A table which shows a set of variables and the number of times each variable occurs (its **frequency**) is called a **frequency table** or **frequency distribution table**.

If a large amount of quantitative data has been collected, it is generally convenient to record the information in a more compact form by combining variables into **groups** or **classes**. Continuous variables, such as time, length, speed, will normally be grouped before the information is collected.

Suppose the leisure centre survey is extended to find the amount of time people spend in the centre.

First the size of each class is decided (say 15 minutes).

Then a table is drawn up of all the classes.

The time each person in the survey has spent in the centre is tallied against the appropriate class and hence the frequencies are found.

Time spent (minutes)	Tally	Frequency
Less than 15	JHT I	6
15–29	JHT JHT	10
30–44	JHT JHT III	13
45–59	JHT JHT II	12
60–74	JHT JHT JHT I	16
75–89	JHT JHT JHT JHT	20
90–104	JHT JHT JHT JHT I	21
105–119	JHT JHT JHT II	17
120–134	JHT JHT IIII	14

1 During a survey to find how knowledgeable the general public is about art, 40 people were asked to name as many artists as possible in one minute. The responses were:

```
 1   5  3  5  1   8  15  1
 2   2  1  3  4   4   1  2
13  11  8  6  1   2   5  2
 3   4  9  2  3  10   1  6
 2   7  1  4  6   4   5  3
```

Use a tally chart to draw up a frequency table for this data.

2 A textile mill spins yarn. The thickness of the yarn is measured at intervals, and the measurements, in millimetres, of a sample of 50 are given below.

```
0.72  0.98  0.81  0.96  0.91  0.90  0.76
0.92  0.95  0.91  0.83  0.91  0.89  0.86
0.93  0.94  0.78  0.93  0.83  0.86  0.91
0.78  0.92  0.88  1.03  1.04  1.01  0.94
1.03  0.90  0.85  0.85  0.91  0.82  0.88
0.95  1.02  0.99  0.97  0.92  0.82  0.90
1.03  0.93  0.94  0.96  0.87  0.93  0.89
0.92
```

Using intervals of 0.70–0.74, 0.75–0.79, 0.80–0.84, etc., draw up a tally chart to obtain the frequency distribution.

3 A small business carried out a survey to find the number of days absence of the employees over one year. The results were:

```
 3   1   4   2   1  15  20   5  15
 0  17  26   0  11   1   3  10   8
15  10  17  10   6  13  12  10  14
 5   8   3  21   0   3  18   3  18
 3  42   9  18  10  21  10   5   6
14   1   5   5   0   5   7  30   9
 0   5   6  25  23   6   4  11  12
```

Collect the information on a frequency table using intervals 0–4, 5–9, 10–14, 15–19, etc.

4 During a survey into changes in the conditions of work of clerical staff, 50 workers gave their present salaries (in £) as:

```
14030  12670  10180  11320   9870
10120  10130  15460  13680   9920
13830  11610  11880  14280  12200
11020  11570  10990   9700  11810
10880  11370  12090   9800   9670
12230  11680   8590   9680  10280
10420  12120   9330  10540   7490
 9240   8990   7630  11010   9180
 8320   8640  15200   8680  12040
 8680   7480   7720   8290   8470
```

Organize the data on a frequency table using intervals £7000–£7999, £8000–£8999, etc.

5 The staff in a medical practice monitored the waiting times of patients from the time the patient sat down until called to see the doctor. The times, in minutes and seconds, were:

```
10:03  12:05   7:15   9:44  11:15
10:02  14:23  12:15  12:42  15:00
 5:43   9:08   9:53   9:03  14:21
 7:24  10:57  12:26   7:13  15:30
10:53  12:26  10:57  13:48   8:00
 9:24  12:48  10:17  11:02   7:48
 9:56  14:09   7:23   9:03   9:59
 9:32   8:05   7:53  14:23  13:03
```

a Write each waiting time to the nearest minute and use a tally chart to obtain the frequencies.

b Summarise the given waiting times in a frequency table using intervals $5.00 \leqslant T < 7.00$, $7.00 \leqslant T < 9.00$, etc., where T is the waiting time.

6 The weights (in kg) of 63 male patients admitted to a ward were recorded as:

```
72.4  68.2  69.3  71.1  66.8  67.2  65.4
68.0  78.9  76.0  70.8  64.3  82.3  70.2
74.2  76.7  65.5  71.6  74.1  68.7  66.8
83.3  74.9  71.5  75.7  71.6  73.2  82.5
73.8  78.2  65.6  76.9  76.8  81.5  77.2
75.8  75.4  80.3  78.0  68.3  76.0  78.5
78.8  71.7  74.4  69.8  77.6  73.4  77.3
74.9  72.4  66.9  73.7  74.4  68.8  82.6
73.7  79.8  74.0  71.8  73.4  76.0  79.2
```

Using intervals 60–, 65–, 70–, etc., summarise the information in a frequency table.

7 An increasing number of couples are choosing to celebrate their wedding in an exotic location. A survey to find the most popular destinations produced the following data (A = Antigua, B = Barbados, F = Florida, J = Jamaica, K = Kenya, L = St Lucia, M = Mauritius, S = Seychelles):

```
B  J  L  S  S  L  K  A  K  B
M  A  A  J  L  L  L  L  B  S
K  J  F  F  B  L  K  K  L  A
M  K  J  J  K  L  B  S  J  K
L  B  J  K  K  L  M  A  L  J
```

Record the data on a frequency table.

8 The number of unoccupied seats on 80 transatlantic flights in one day were:

```
32   8   9   6  12  30   9  11   5  39
 6  25  26  42  33  16  13  30   5  29
43  34  11  26   2  39  35  19  20  40
15  11  20  34  31  17  23   2  17  15
32   3  44   6   1   7  26  35  18  25
37   4  39  37  34  26  33   7  21  16
18  15  29  35  21   6  40  39  13  12
 4   4  38  39  12   0   4  33  34  18
```

Summarise the information on a frequency table using class intervals 0–4, 5–9, 10–14, 15–19, etc.

9 The following data are the weights (w), in kilograms, of the luggage of 50 passengers boarding a charter flight to Europe. The luggage allowance was 20 kg per person.

```
18.04  22.32  18.02  20.16  20.50  15.18
13.48  19.90  17.98  19.76  17.12  22.00
21.44  20.04  17.30  16.30  20.24  18.76
17.24  19.96  16.72  15.92  19.66  17.94
18.70  19.94  15.02  17.58  16.80  17.52
17.10  19.82  22.44  15.80  19.02  18.82
15.76  19.46  18.29  20.23  17.78  15.82
15.90  16.46  16.72  18.22  19.00  20.50
17.62  20.44
```

Group the information into classes of width 2 kg using intervals $12.00 < w \leqslant 14.00$, $14.00 < w \leqslant 16.00$, $16.00 < w \leqslant 18.00$ etc. and display in a frequency table.

10 The number of faults found in a sample of 50 micro chips was:

```
1  0  0  0  2  0  0  0  1  0
0  0  1  1  0  1  0  0  2  0
0  0  0  0  1  0  0  0  0  0
2  0  0  1  0  1  1  0  3  0
1  1  0  0  2  0  2  1  0  1
```

Summarise the data on a frequency table.

11 A firm manufactures ball bearings for the motor and motorbike industries. In order to monitor accuracy, samples are taken at intervals from five machines and the diameters (d) of those bearings are measured in millimetres.

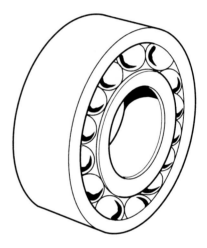

The results of one set of samples are given below:

```
49.46  50.15  51.15  51.36  49.72
50.41  50.03  48.16  50.26  49.43
49.66  49.31  50.32  49.76  46.78
50.16  50.98  51.15  49.40  48.88
48.90  50.47  49.33  50.08  48.20
49.76  49.67  50.05  50.70  49.46
50.14  49.39  52.63  49.93  49.19
50.09  49.27  49.63  51.09  48.21
```

Summarise the data on a frequency table using class intervals of 1 mm:

$46.00 \leqslant d < 47.00$, $47.00 \leqslant d < 48.00$, ..., $52.00 \leqslant d < 53.00$.

08 Statistics on Display

8.1 Pictorial representation of data

The presentation of data in the form of tables has been considered in Unit 7. However, most people find that the presentation of data is more effective, and easier to understand, if the data is presented in a pictorial or diagrammatic form.

The pictorial presentation used must enable the data to be more effectively displayed and more easily understood. The diagrams must be fully labelled, clear and should not be capable of visual misrepresentation. Types of pictorial representation in common use are the pictogram, bar chart, pie chart and frequency polygon.

Statistical packages may also be used to present data in a variety of ways. You cannot, however, rely completely on a computer to produce your pie charts, pictographs, etc. You must also be able to carry out the necessary calculations yourself and draw the most appropriate diagrams for the given data.

There are many ways of presenting data in pictorial form. It is clearly necessary to be able to interpret correctly any diagrams given.

The general interpretation of statistical pictures and graphs is that the bigger the representation, the larger the population in that group. However, it is also possible to interpret statistical diagrams so as to be able to calculate the population of each group.

8.2 Pictograms

In a **pictogram** data is represented by the repeated use of a pictorial symbol. The example below shows how a pictogram works.

EXAMPLE

A survey of 1000 people living in Freeton was taken, to see what colour of cars they owned. Represent this data in the form of a pictogram. The results of the survey were:
Colour of cars Number of cars

Colour of cars	Number of cars
Red	60
White	200
Blue	100
Grey	80
Gold	50
Black	30

Key: 🚗 = 20 cars

Here is one possibility. A full car symbol represents 20 cars; half a car represents 10 cars. It is not possible to show small fractions of a symbol accurately, and the detail required should not normally be to more than half of a symbol (but certain symbols may allow for a quarter).

EXERCISE 8.1

1   The numbers of bottles of champagne sold in five villages is shown on the following pictogram:

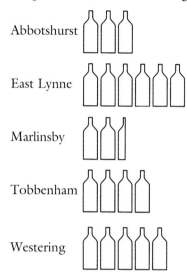

Abbotshurst

East Lynne

Marlinsby

Tobbenham

Westering

a In which village were the most bottles sold?
b How many bottles were sold in Martinsby?
c How many bottles were sold in total?

2   The contents of a fruit bowl comprised:

Apples	7	Bananas	3
Pears	5	Peaches	7
Kiwi fruit	6	Oranges	2

Illustrate this data by means of a pictogram.

3   Students in a department of a college were asked about the type of accommodation in which they lived. The data was:

Flat	25	Semi-detached house	40
Maisonette	5	Detached house	30

Illustrate this data by means of a pictogram.

4   In a survey to find the most popular design on Christmas cards, 600 people were asked which animal they preferred. The results are shown on the pictogram.

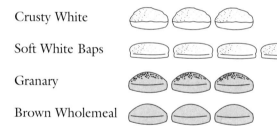

Reindeer

Penguin

Robin

Polar bear

(A full picture represents 50 voters.)

a How many people chose the polar bears design?
b What percentage of people chose the reindeer?
c Find the ratio of the votes for robins to the votes for penguins.

5   The number of people present in a Paris fashion show were:

Individual buyers	54
Store buyers	18
Celebrities	27
Photographers	45
Journalists	36

Illustrate this data by means of a pictogram.

6   The numbers of employees in four solicitors' offices were:

Archibald and Archibald	8
Dugdale, Wynne and Luff	10
JSC Weston-Hough	6
Mordecai and Sons	12

Draw a pictogram to represent this data.

7   A college canteen carried out a survey to decide which type of bread roll to serve. The answers are shown in the pictogram.

Crusty White

Soft White Baps

Granary

Brown Wholemeal

(One roll represents 4 votes.)

a Which was the favourite type of roll?
b How many customers preferred wholemeal rolls?
c What percentage preferred white bread?

8   The number of patients on a register of six doctors in a group practice is:

Dr Smith	2400
Dr Rawlings	1800
Dr Wong	3000
Dr Payne	2700
Dr Williams	23100
Dr Fisher	1500

Show this information on a pictogram.

9 The number of residents in five rest-homes is shown on the pictogram.

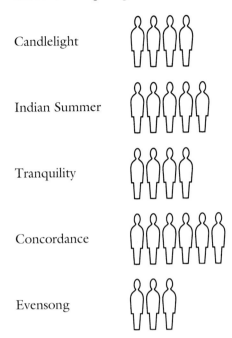

Candlelight

Indian Summer

Tranquility

Concordance

Evensong

(Each picture represents five residents.)

a How many residents are there in Candlelight?

b Which rest-home has the smallest number of residents?

c What is the total number of residents?

10 The number of flights for each airline out of Gatwick in a one-hour period was:

British Airways	8	Aer Lingus	2
Swissair	1	Britannia	6
Virgin Atlantic	1	Monarch	3

Illustrate this data on a pictogram.

11 One clock represents one hour on the following pictogram, which shows the time taken to cross the English Channel by different routes.

Dover–Calais

Dover–Ostend

Portsmouth–Caen

Portsmouth–Cherbourg

Plymouth–Roscoff

a Which route takes the longest time?

b How long does it take to go from Portsmouth to Cherbourg?

c What is the difference in the time taken to cross the Channel between the Dover-to-Calais route and the Dover-to-Ostend route?

12 The management of a car plant wanted to know how many of the workers used the cars produced by their company. They decided to carry out a survey of the cars parked in the factory's car park one day. The results were:

Ford	35	Citroen	20
Rover	30	Renault	10
BMW	5	Vauxhall	25

Construct a pictogram to illustrate the results.

13 The workforce of a factory were asked by which mode of transport they came to work. The results are shown in the following pictogram, each figure representing 10 workers:

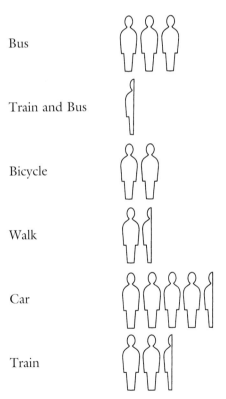

Bus

Train and Bus

Bicycle

Walk

Car

Train

a How many of the workers walk to work?

b Which form of transport is used least to get to work?

c What is the total work force of the factory?

8.3   Bar charts

A **bar chart** is a diagram consisting of columns (i.e. bars), the heights of which indicate the frequencies. Bar charts may be used to display discrete or qualitative data.

EXAMPLE

Fifty households were surveyed, and the number of children in each family was recorded as follows:

Children in family	Frequency
0	8
1	11
2	17
3	8
4	5
5	1

Represent this data by means of a bar chart.

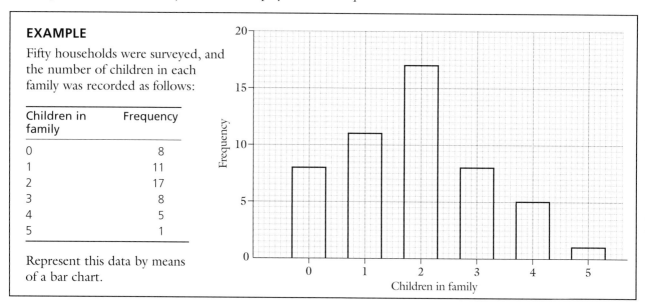

Dual bar charts

Dual bar charts are used when two different sets of information are given on connected topics.

EXAMPLE

The number of people over 17 years old, and the number of people holding driving licences in a particular street were found over a period of years.
These are as shown below.
Represent this data by means of a dual bar chart.

Year	1995	1996	1997	1998	1999	2000
No. of people over 17	32	27	29	31	33	39
No. of people with driving licence	12	17	19	11	24	28

Sectional bar charts

Sectional bar charts, or **component bar charts**, are used when two, or more, different sets of information are given on the same topics. They are particularly useful when the *total* of the two or more bars is also of interest.

EXAMPLE

The numbers of saloons and hatchbacks sold by a garage were recorded.

Month	Jan	Feb	Mar	Apr	May	Jun
Saloons	18	7	8	12	10	13
Hatchbacks	16	12	9	7	9	8

Represent this data by means of a sectional bar chart.

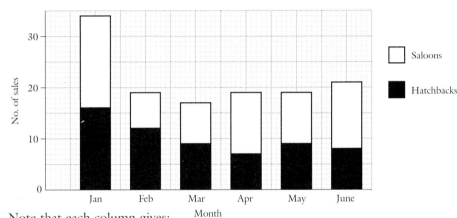

Note that each column gives:

(i) the number of saloons sold,

(ii) the number of hatchbacks sold, and

(iii) the total number of cars sold during that month.

All three sets of information can rapidly be compared by using the same diagram.

EXERCISE 8.2

1   The bar chart shows the type of trees recorded during a survey of a section of a forest.

Oak
Elm
Chestnut
Beech
Conifer
Cedar

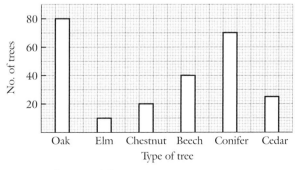

a Which tree was seen most frequently?

b Reconstruct the frequency table.

c What was the total number of trees growing in this area?

2 Twenty people noted the television channel they were watching at 8.15pm on two successive nights. The results were:

	First night	Second night
BBC 1	6	7
BBC 2	4	1
ITV	6	6
CHANNEL 4	3	4
SATELLITE	1	2

Draw a suitable bar chart to illustrate this data.

3 Thirty students designed ball gowns. The colours of the gowns were 12 black, 5 gold, 10 red, 2 blue, and 1 green.

Illustrate this information by means of a bar chart.

4 A graphics company investigated how many hours their employees actually worked on a computer during a week in January 1998 and again in 1999. The results were:

Time per week (hours):	0	0–1	1–2	2–4	4–10	10–20	20–40
1998	32	17	4	10	2	1	25
1999	17	2	5	11	12	21	34

Represent this data by means of a dual bar chart.

5 The number of students in an art class on six successive evenings were:

	Eve 1	Eve 2	Eve 3	Eve 4	Eve 5	Eve 6
Male	11	9	8	9	7	6
Female	7	8	9	10	11	13

Illustrate this information by means of a section bar chart.

6 In a three-month period, the number of days in which different products were advertised on two hoardings were compared. These are shown in the dual bar charts below.

a One hoarding was in an inner city, and the other one was in a suburban area.
   Which hoarding was in the inner city?

b How many more days did Hoarding A advertise alcohol than Hoarding B?

c Which products were advertised only on Hoarding A?

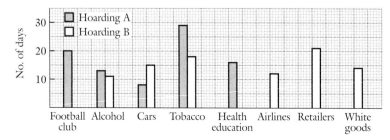

7 The numbers of large appliances sold by an electrical retailer during one day were:

Washing machines   35
Cookers   28
Televisions   38
Video Recorders   12
Refrigerators   21

Illustrate this data by means of a bar chart.

8 An insurance company keeps records of life endowment policies sold by its representatives. In a six-month period the number of policies sold by their top salesmen were:

	With Profits	Without Profits
January	11	21
February	15	12
March	28	13
April	21	20
May	18	27
June	16	30

Represent this data by means of a section bar chart.

9 The number of houses sold by four agents in the first six months of 1998 and 1999 were:

	1998	1999
John	27	29
Mary	15	24
Carl	28	19
Latha	22	32

Represent this data by means of a dual bar chart.

10 Peter and Frances Mead decide to apply for a franchise. They investigate the possible companies selling fast food and the results of their investigations are shown on the bar chart below.

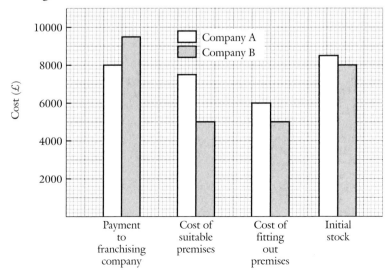

    a What is the total cost of franchising with Company A?

    b What is the total cost for franchising with Company B?

    c How much greater is the payment to franchising Company A than to Company B?

    d If Peter and Frances franchise with Company A, they would expect to have double the profits than if they had franchised with Company B. What other information would they need to know before they chose which company to franchise with?

11 The numbers of patients in six wards of a general hospital were:

Children	32	Cardiac	16
Orthopaedic	17	Surgical	28
Gynaecological	24	Geriatric	36

Represent this data by means of a bar chart.

12 The numbers of residents, and the numbers of staff employed at nursing homes in a small town were:

Nursing home	Residents	Staff
Golden Memories	45	28
Peacehaven	21	16
Autumn Leaves	41	24
Silver Threads	31	19

Represent this data by means of a sectional bar chart.

13 The birth rate per 1000 of population and the infant mortality rate per 1000 live births were found for a number of countries:

	The Gambia	Hungary	Italy	Samoa	UK
Birth rate	47.5	12.2	10.1	39.1	13.3
Infant mortality rate	174	20.4	10.9	4.4	9.4

Construct a dual bar chart to represent this data.

14 The goals scored in 42 football league matches on Saturday 26 March 2000 were:

Number of goals in match	0	1	2	3	4	5	6
Number of matches	1	9	10	12	6	3	1

Illustrate this information by means of a bar chart.

15 Jean-Paul and Michelle noted how the cost of their summer holiday had varied in the last two years. The money had been spent as shown:

	Travel	Rent of Villa	Food	Drink	Entertainment	Insurance
Cost 1998 (£)	720	820	550	140	250	110
Cost 1999 (£)	650	920	480	160	150	120

Represent this data by means of a dual bar chart.

16   The percentage of trade of certain countries with the UK and the USA
was calculated in 1996 to be as follows:

	Barbados	Canada	Hong Kong	Ireland	Luxembourg	Sweden
% trade with E.C.	16.8	1.4	5.0	25.7	3.1	9.6
% trade with USA	13.5	82.3	25.4	8.2	6.3	8.3

Illustrate this information by means of a dual bar chart.

17   The number of employees in five small printing companies were:

Firm A   6 secretaries   and 17 other staff
Firm B   8 secretaries   and 14 other staff
Firm C   11 secretaries   and 10 other staff
Firm D   5 secretaries   and 12 other staff
Firm E   7 secretaries   and 6 other staff

Represent this data by means of a sectional bar chart.

8.4   Pie charts

A **pie chart** is another type of diagram for displaying information. It is particularly
suitable if you want to illustrate how a population is divided up into different parts
and what proportion of the whole each part represents. The bigger the
proportion, the bigger the slice (or 'sector').

EXAMPLE

Represent by a pie chart the following data.

The mode of transport of 90 students
into college was found to be:

Walking	12
Cycling	8
Bus	26
Train	33
Car	11
Total	**90**

Represent this data by means of a pie chart.

A circle has 360°. Divide this by 90 to give 4°. This is then the
angle of the pie chart that represents each individual person.

Since 12 people walk to college, they will be
represented by $12 \times 4° = 48.°$

Similarly for the others:

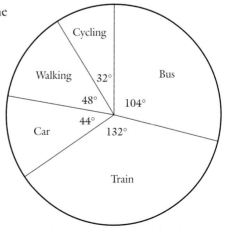

	Angle in pie chart
Walking	$12 \times 4 = 48°$
Cycling	$8 \times 4 = 32°$
Bus	$26 \times 4 = 104°$
Train	$33 \times 4 = 132°$
Car	$11 \times 4 = 44°$
	Total $= 360°$

Method for calculating the angles on a pie chart

Here is a summary of how to work out the size of each bite of 'pie'.

(i) Add up the frequencies. This will give you the total population (call it p) to be represented by the pie.

(ii) Divide this number into 360.

(iii) Multiply each individual frequency by this result. This will give you the angle for each section of the pie chart.

Interpreting pie charts

The initial interpretation is the fact that the largest portion of a pie chart relates to the largest group, and the smallest portion to the smallest group. However, if any of the data is known, the rest of the data can be calculated.

EXAMPLE 1

The pie chart below shows the number of students in different sections of a college. 220 students are in the Construction department.

a How many students are there in the college?

b How many students are there in Catering?

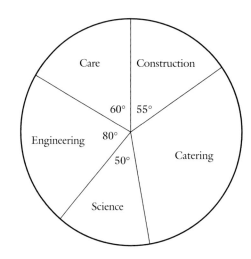

a 55° represents 220 students.

$\therefore$ 1° represents $\dfrac{220}{55} = 4$ students.

The complete circle (360°) represents $4 \times 360 = 1440$ students.

$\therefore$ There are 1440 students in the college.

b The angle representing Catering is
$360 - (80 + 55 + 60 + 50) = 115°$.

$\therefore$ The number of students in Catering is $4 \times 115 = 460$.

EXERCISE 8.3

Illustrate the data given in questions 1, 2 and 3 below by means of a pie chart.

1 The types of central heating used by households in a village were:

Solid fuel	14
Gas	105
Electricity	41
None	20

2 240 students were asked what they were intending to do during next year. The results were:

80 going to university
86 staying at college
64 going into employment
10 with no firm intention.

3 The numbers of bedrooms in 720 houses recorded as:

1 bedroom	80
2 bedrooms	235
3 bedrooms	364
4 bedrooms	39
5 bedrooms	2

4 The pie chart shows the different drinks sold at lunchtime in a college. 720 drinks were sold in total.

Find the number of each different drink sold during lunchtime:

a Coke
b Orange
c Coffee
d Chocolate

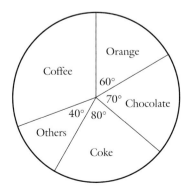

5 The number of special birthday cards sold by a newsagents in a week were:

Mum	42
Dad	34
Grandad	18
Granny	16
Brother	25
Sister	17
Son	15
Daughter	13

Illustrate this information on a pie chart.

6 An artist designed book jackets for 225 books during a five-year period with one publishing house. The books were classified as Thriller, Romance, Travel, Hobby and Science.

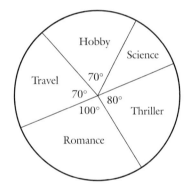

a What size angle represents science books?
b How many jacket designs did the artist create for science books?
c How many jacket designs were created for thrillers?
d What fraction of the designs were for romances? (Give your answer in its lowest terms.)

7 Each pound spent at the Winchester Theatre Royal box-office is used to meet the theatre's expenses as follows:

Performance fees	60p
Salaries	17p
Premises and depreciation	10p
Administration	5p
Publicity	5p
Equipment	3p

Draw a pie chart to show how each pound is spent.

8 The number of clients of each of the partners in a business were:

Kerry 43
Michell 22
Owen 18
Richard 7

Illustrate this information on a pie chart.

9 The proportion of the cost of manufacturing a dinner service was:

Raw material 5%
Manufacture cost 28%
Hand painting cost 51%
Overheads 5%
Profit 11%

Represent this data on a pie chart.

10 The pie chart shows the different types of petrol which a garage sold in one week.
The garage sold 25 000 gallons of diesel.

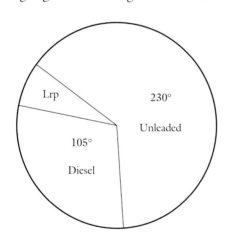

a How much unleaded petrol was sold?
b How much lead replacement petrol was sold?
c What were the total sales?

11 A health centre calculated the distance patients from one practice lived from their GP's surgery:

Under 1 mile: 510 patients
Between 1 and 2 miles: 1230 patients
Between 2 and 3 miles: 140 patients
Over 3 miles: 280 patients.

Illustrate this information on a pie chart.

12 The pie chart shows the type of dwellings in which people in a village live. There are 720 dwellings in the village. By measuring the angles, find how many are:

a detached houses
b bungalows
c semi-detached houses

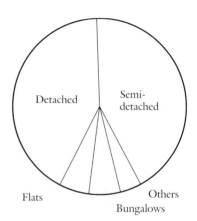

13 The number of villagers undertaking an activity to ensure 'active living' were:

Swimming 1020
Walking 3580
Jogging 105
Keep fit centre 85
Other 610

Draw a pie chart to illustrate this data.

14 The holiday destinations of 60 people were:

France	Spain	Greece	Tunisia	USA	Caribbean	Portugal
21	15	6	3	8	5	2

Represent this information by means of a pie chart.

15 At a sports centre, the ages of 100 people were recorded as follows:

Under 20 years   30
20 to 29 years   15
30 to 39 years   12
40 to 59 years   14
60 years and over   29

Construct a pie chart to illustrate this data.

16 The pie chart shows the number of passengers flying from London to Miami on one afternoon. 1800 passengers in total flew this route on that afternoon.

Find the number flying:

a Virgin

b American Airlines

c British Airways

The plane used by Virgin is a Boeing 747 seating 370 passengers.

d What percentage of the Virgin seats were occupied?

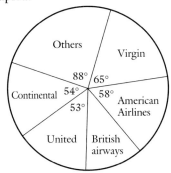

17 Out of 90 employees in a manufacturing company, there were 10 managers, 15 salesmen, 52 production line workers, 4 typists, and 9 quality controllers.
Illustrate this information on a pie chart.

18 A manufacturer of combined harvesters commissioned a survey on the use of agricultural land in South Australia. The result is shown in the pie chart.

a Find the percentage of land use for wheat.

b The total acreage is 3 000 000 acres. What area was used for hay?

19 In 1984, in the UK, there were 4 500 000 people employed in manufacturing. Of these, 600 000 were in the food industry, 659 000 were in mechanical engineering, 500 000 in electronic engineering and 400 000 in printing. Identify the remainder as 'other industries', and draw a pie chart to represent this information.

8.5   Comparative pie charts

When we have carried out two similar surveys, we can represent the results by drawing two separate pie charts. For example, the two pie charts below show the results of a survey of two campsites, one of which has 25 male campers and 15 female campers, while the second campsite has 155 male campers and 205 females.

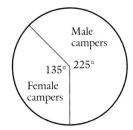 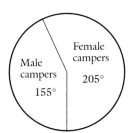

However, although these pie charts show the information for each campsite, they do not show that one campsite has far more people in it than the other. To show this information more accurately, we draw pie charts of different sizes.

As the area of a circle of radius r is πr^2, the ratio of the area of the two pie charts is equal to the ratio of the **square** of the radii of the two pie charts.

Campsite A has 40 campers, while Campsite B has 360.

To find the ratio in its simplest form, divide the larger number by the smaller number:

$$\frac{360}{40} = 9$$

Hence the number of campers, which are in the ratio $40 : 360$, are in the ratio of $1 : 9$.

If we draw two circles, one of radius r cm and the other of radius $3r$ cm, the areas of those circles will be in the ratio $1 : 9$.

Using 2 cm and 6 cm for the radii of the two circles, we can construct the two pie charts below.

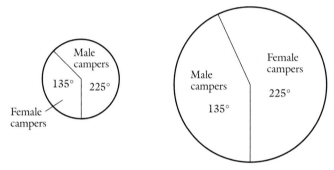

These two pie charts now show both:
  (i) information on the proportion of male and female campers at each camp site, and
 (ii) information on the total number of people camping at each camp site.

EXAMPLE

Two florists, Blooms Galore and Floral Fantasy sell tulips for Mothering Sunday. Show their sales, detailed below, by means of two comparative pie charts.

	Blooms Galore	Floral Fantasy
Yellow tulips	250	9600
Red tulips	470	1920

The total sales are: 720 by Blooms Galore
                   11 520 by Floral Fantasy

Remember that to find the ratio in its simplest form, we divide the larger number by the smaller number:

$$\frac{11\,520}{720} = 16$$

The ratio of the sales are $720 : 11\,520$ or $1 : 16$
The ratio of the radii of the circles is $1 : \sqrt{16}$ or $1 : 4$

Therefore, we draw two circles, one with radius 1 cm and the second with radius 4 cm.

The angles within each pie chart are:

Blooms Galore	Yellow 125°	Red 235°
Floral Fantasy	Yellow 300°	Red 60°

The required pie charts are:

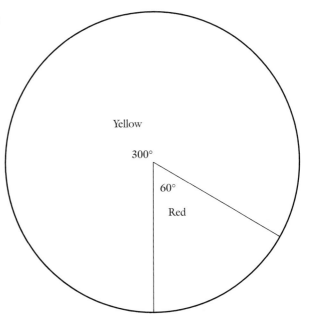

---

1   Two fishing boats return to port with a mixture of cod and herring. The catches are:

"Lucky Lady"
    17 tons of cod and 13 tons of mackerel
"Fair Winds"
        42 tons of cod and 78 tons of mackerel

Show this information by means of two comparative pie charts.

2   The numbers of cars and lorries passing abridge on a motorway were noted between 5 pm and 6 pm on Friday, and between 4 am and 5 am on Monday.

The numbers of vehicles were:

Friday   970 cars and 110 lorries
Monday   48 cars and   72 lorries

Show this information by means of two comparative pie charts.

3   The number of holidaymakers in two hotels on one day in July was noted as:

	Adults (male)	Adults (female)	Children
Sea View Hotel	15	14	7
Solent Hotel	210	232	134

Show this information by means of two comparative pie charts.

4   The holiday destinations of 540 people were as follows:

France	Spain	Greece	Tunisia	USA	Caribbean	Portugal
186	180	2	12	33	75	33

Use two comparative pie charts to compare this data with the data in Question 14 of Exercise 8.3.

8.6   Line graphs

A bar chart can be replaced by a line graph, provided that the quantity on the horizontal axis is continuous, e.g. age, temperature or time.

In this case the data is plotted as a series of points which are joined by straight lines.

Line graphs associated with time are called **time-series graphs**. They are used, for example, by geographers to illustrate monthly rainfall or yearly crop yield, etc., and by businesses to display information about profits or production over a period of time.

They show trends and have the advantage that they can be easily extended.

EXAMPLE

The temperatures, recorded every six hours, of a patient in a hospital ward are given on the table:

	Mon.			Tues.					Wed.
Time (hours)	06	12	18	00	06	12	18	00	06	12
Temperature (°F)	99.0	99.12	99.12	99.2	99.2	98.99	98.68	98.6	98.6	

Represent this data by means of a line graph.

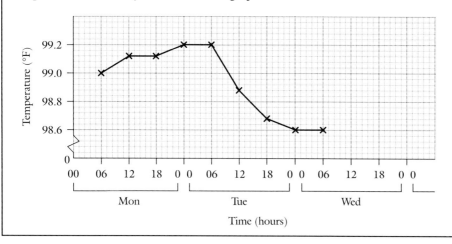

Illustrate the data given in questions 1, 2 and 3, using a line graph.

1 The rainfall during a period of six months was:

Month	Jan	Feb	Mar	Apr	May	Jun
Rainfall (mm)	75	192	86	89	25	19

2 The maximum temperatures for six successive months at Sunbourne were:

Month	Apr	May	Jun	Jul	Aug	Sep
Temperature (°C)	61	74	72	91	85	56

3 The girth of a tree was:

Age (years)	10	20	30	40	50	60
Girth (cm)	25	63	98	135	170	210

4 In an art class, students were asked to draw a bowl of fruit. The number of grapes drawn by the students were:

Number of grapes	0	1	2	3	4	5
Number of students	2	7	6	3	2	1

Construct a line to show this information.

5   The graph below shows the number of births (measured along a vertical axis) in a given year (measured along a horizontal axis).
Answer the questions below by reading off the values from the graph.

a   Estimate the number of children born in 1915.

b   In which years were approximately 825 000 children born?

c   Which year had the lowest number of births?

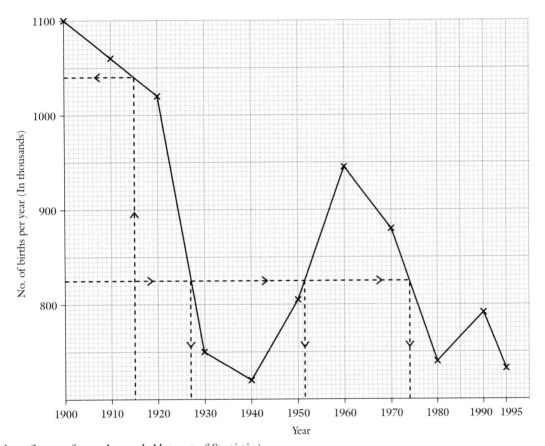

(Based on figures from *Annual Abstract of Statistics*)

6   Mordecai makes cuddly toys. The number of koala he made in one week is shown on the line graph below:

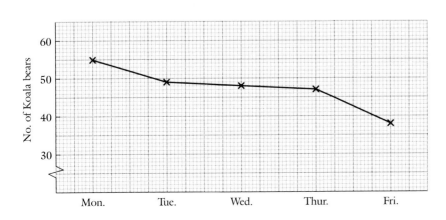

a The trend of this graph is downwards. Give a possible reason for this.

b During the following week, Mordecai's production of koala bears was:

Mon	Tue	Wed	Thur	Fri
41	46	49	48	42

What was the total production for each week?

7 The graph shows the median gross weekly earnings of women from 1976 to 1994.

a Estimate the median weekly earnings in 1981.

b Estimate the median weekly earnings in 1993.

c Why is the answer to **b** a better estimate than the answer to **a**?

d In which year did the average weekly wage reach £200?

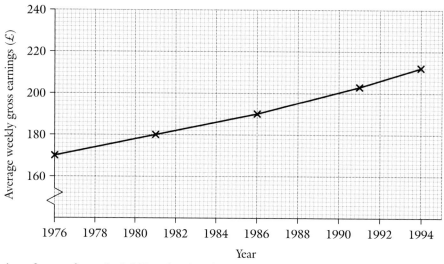

(Based on figures from *Social Trends*, 1998.)

8 The numbers of passengers carried by an airline (in thousands) were as follows (Sp = spring, Su = summer, etc.):

	1997				1998				1999		
Sp	Su	Au	Wi	Sp	Su	Au	Wi	Sp	Su	Au	Wi
21	48	31	17	22	49	29	18	23	41	24	25

a Plot this information on a line graph.

b State any trends which the data suggests.

9 The numbers (in thousands) of steel pipes made by a company were:

Jan	Feb	Mar	Apr	May
24	23	18	14	27

Illustrate this information by means of a line graph.

10 The graph below shows the number of cars supplied per month by a car manufacturer to a garage's sales section:

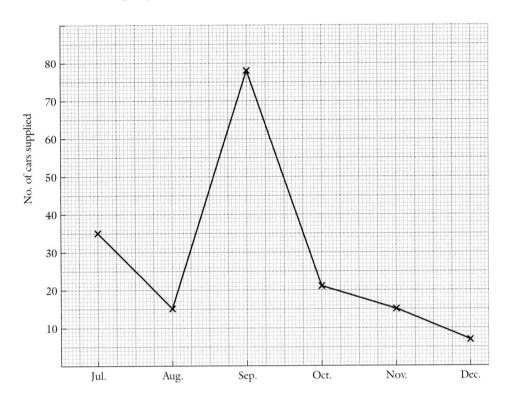

a In which month were most cars supplied?
   Give a possible reason for this large amount.

b How many cars were supplied in November?

c What was the total number of cars supplied during the six months?

d What was the average number of cars supplied per month during this period?

8.7 Histograms

In Section 8.3 we used bar charts to show the frequencies of qualitative data and also of quantitative data which had only a few discrete values (i.e. the number of children in a family).

When discrete data has a greater range of values, or when data is continuous, it is usually helpful to group the data into classes and show the frequencies of the classes.

When we do this with discrete data, we are actually treating it as if it were continuous. For example, the discrete examination marks out of 60 gained by 100 students may be grouped in tens as follows:

Marks	0–	10–	20–	30–	40–	50–60
Frequency	3	11	21	44	15	6

For discrete data, we should not write the groups as 0–10, 10–20, 20–30, 30–40, 40–50, and 50–60 as it is then not possible to know to which groups the numbers 10, 20, 30, 40 and 50 have been assigned. With continuous data, this does not cause a problem as the probability of obtaining a value of exactly 10 is zero (measurements will always be slightly above or below).

The diagram which shows frequencies of data grouped like this is called a **histogram**.

Since the data is continuous, or treated as such:

(i) the histogram must have a continuous horizontal scale;

(ii) each column will have, as its base, its class interval;

(iii) there will be no spaces between the columns, unless the frequency relating to a class interval is zero.

A histogram constructed from discrete data

EXAMPLE

The students' examination marks were grouped into tens:

Marks	0–	10–	20–	30–	40–	50–60
Frequency	3	11	21	44	15	6

The histogram is as shown below:

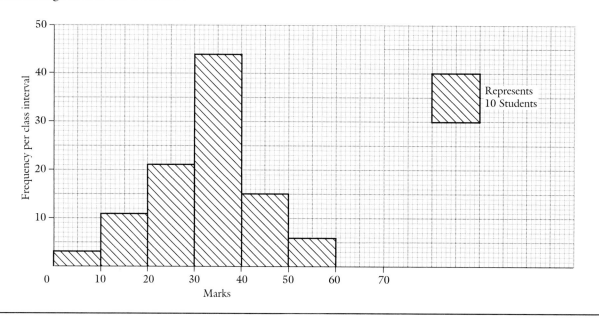

When the class intervals are all equal, the heights of the columns (being proportional to the areas) show the frequencies directly.

Because this is often called a bar chart or a frequency diagram, as well as the correct name, histogram, it is common to label the vertical axis simply 'frequency' when the class intervals are all equal. More precisely, the vertical axis should be marked 'frequency per class interval'.

A histogram constructed from continuous data

EXAMPLE

The diameters of 140 apples in a box were measured and the results were recorded as follows:

Diameter (cm)	4–	5–	6–	7–	8–9
Frequency	12	20	56	40	12

This frequency table tells us that the apples have been put into groups or classes of 1 cm width.

In the first class there are 12 apples whose diameters might have any measurement from 4 cm up to, but not including, 5 cm.

If the apples were being measured on the millimetre scale and an apple was measured and found to have a diameter of exactly 5.0 cm, it would belong to the second class.

The histogram representing this data is as shown below.

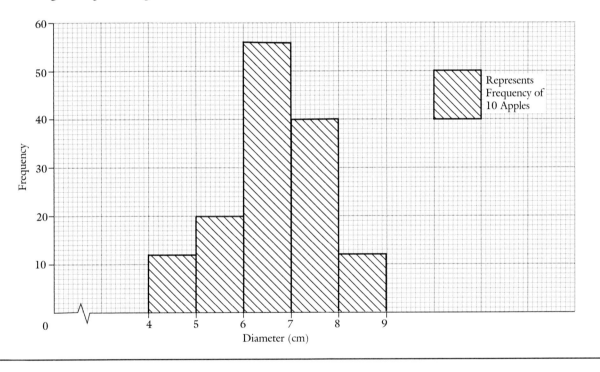

Class boundaries

If the class intervals are defined by rounded numbers as in 0–9, 10–19, 20–29, 30–39 etc. and the data is continuous, we need to be able to allocate values such as 9.3, 29.5 and 29.8 into the groups.

We define boundaries to the classes at the half-way points of 9.5, 19.5, 29.5, 39.5, etc. Values below the boundaries belong to the class below, and values at and above the boundaries belong to the class above.

The **upper class boundary** is the maximum possible value which would be in that class.

The **lower class boundary** is the minimum value which would be in that class.

For example, 9.3 belongs in the class 0–9 and 29.5 and 29.8 belong in the class 30–39.

The class boundaries will depend on the degree of rounding. For example:

(i) The heights of trees may be measured to the nearest metre. The class 4–7 m contains trees of height 3.5 m up to (but not including) 7.5 m and the class boundaries are 3.5 and 7.5 m.

(ii) The heights of buildings may be measured more accurately, to the nearest 0.1 m. The class 4.0–7.0 m contains buildings of height 3.95 m up to (but not including) 7.05 m and has class boundaries 3.95 m and 7.05 m.

Note. Age boundaries are different from those for other measures. The class of student ages 17–19 years has class boundaries of 17 and 20 years. This is because students aged 19 are 19 until their 20th birthdays.

Sometimes we are able to show the class boundaries clearly when we draw the columns of the histogram. If possible, we do so. With other data the scale does not allow such fine detail. In this case, it is understood that the difference between the class boundary and the rounded value in the 'frequency table' is too small to show on the graph paper.

A histogram constructed from continuous data showing class boundaries

EXAMPLE

The frequency table for the heights of 108 conifers each measured to the nearest metre is as follows:

Height of conifer (m)	1–3	4–6	7–9	10–12	13–15
Diameter (cm)	12	24	27	30	15

Class boundaries are 0.5, 3.5, 6.5, 9.5, 12.5 and 15.5 (m). Each class width is 3 m, and this is the width of the base of each column in the histogram.

In this case, we are able to show the class boundaries clearly on the histogram.

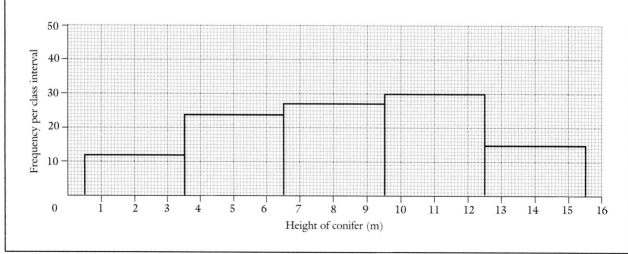

8.8   Histograms with bars of unequal widths

Histograms often have class intervals of different widths. In these cases the **area** of the bar or column represents the frequency, whereas in a bar chart the **height** of the bar represents frequency.

Thus: **in histograms with columns of unequal widths, the heights of the columns are found by dividing the frequency by the width of the class interval**.

The scale on the vertical axis is 'Frequency per unit class interval' or 'Frequency density' where

$$\text{Frequency density} = \frac{\text{Frequency of class interval}}{\text{Width of class interval}}$$

The data in the example on page 60 could be grouped as follows:

Height of conifer (m)	1–2	3–6	7–9	10–11	12–14	15
No. of conifers	6	30	27	28	12	5

Group	1–2	now has class interval	0.5–2.5,	class width = 2
	3–6	now has class interval	2.5–6.5,	class width = 4
	7–9	now has class interval	6.5–9.5,	class width = 3
	10–11	now has class interval	9.5–11.5,	class width = 2
	12–14	now has class interval	11.5–14.5,	class width = 3
	15	now has class interval	14.5–15.5,	class width = 1

The data to be plotted therefore is:

Height of conifer (m)	1–2	3–6	7–9	10–11	12–14	15
Frequency density	3	7.5	9	14	4	5

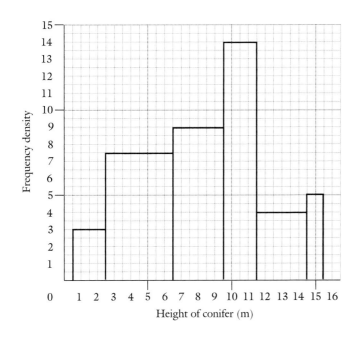

1 The speeds of 100 cars on a motorway were recorded. The data found was:

Speed (mph)	30–40	40–50	50–60	60–70	70–80	80–90
No. of cars	2	11	35	42	9	1

Represent this data by means of a histogram.

2 The heights of 80 students were recorded. The data was:

Height (cm)	150–160	160–170	170–180	180–190	190–200	200–210
No. of students	4	7	15	47	6	1

Represent this data by means of a histogram.

3 Below is a table showing the number of seats not booked on an airline's daily flight between London and Miami over 10 weeks.

19	1	8	11	15	19	21	17	1	23	19	11	12	15
21	11	8	4	15	27	21	20	14	18	7	11	23	21
8	17	1	19	12	16	21	25	28	29	17	15	11	8
16	8	2	8	6	10	11	15	9	8	6	2	3	8
21	18	27	32	37	4	11	19	21	34	21	15	12	11

By means of a tally chart, find the frequencies in the class intervals 0–4, 5–9, 10–14, ..., 30–34, 35–39. Represent this grouped data by means of a histogram.

4 The number of accidents in High Town was recorded over two years, and the information was grouped in weekly periods.

Number of accidents	0–1	2–3	4–5	6–9	10–14	15–17	18–20
Number of weeks	1	7	12	36	28	19	1

Represent this data by means of a histogram.

5 The IQs of 100 students were measured and the results were as follows:

IQ	100-109	110–119	120–129	130–139	140–159
No. of students	12	34	38	13	3

Represent this data by means of a histogram.

6 The wages of 50 workers were as below. Represent this data using a histogram.

Weekly wage (£)	70–80	80–100	100–150	150–175	175–200	200–300
No. of workers	2	8	18	14	5	3

8.9 Frequency polygons

A method of presenting data which is an alternative to a histogram is the **frequency polygon**. They are often used to compare frequency distributions, i.e. to compare the 'shapes' of the histograms, because it is possible to draw more than one frequency polygon on the same graph. It is easier to make comparisons using frequency polygons than using histograms.

For ungrouped data, the frequencies are plotted as points. For grouped data, which is more usual, the frequencies are plotted against the mid-point of the class interval. In both cases the points are joined with straight lines.

EXAMPLE 1

The heights of 80 students were recorded. The data was:

Height (cm)	150–160	160–170	170–180	180–190	190–200	200–210
No. of students	4	7	15	47	6	1

Represent this data by means of a frequency polygon.

The new table is:

Mid-point of class	155	165	175	185	195	205
Frequency density	4	7	15	47	6	1

Frequency polygon

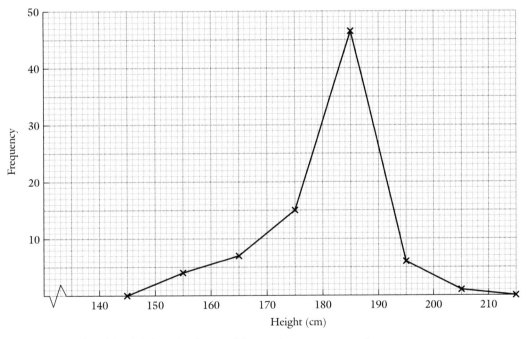

The polygon is completed by joining the first and last points to zero at the mid-point of the next bar.

When frequency polygons are used to compare two sets of data, it is the shapes of the distribution which are important. For this reason, the mid-points of the classes are often plotted against the actual frequencies, rather than the frequency densities which would be used for a histogram.

EXAMPLE 2

The mock examination results in Mathematics for two successive GCSE groups are recorded on the table below.

Mark	1–20	21–40	41–60	61–80	81–100
Group 1 % frequency	5	12	35	28	20
Group 2 % frequency	7	26	48	9	10

a Draw the frequency polygon for each group.

b Assuming the ability of the pupils was the same in each year, comment on the mock examination papers.

a In this example, the percentage frequencies are plotted against the class mid-points, which are 10.5, 30.5, 50.5, 70.5 and 90.5.

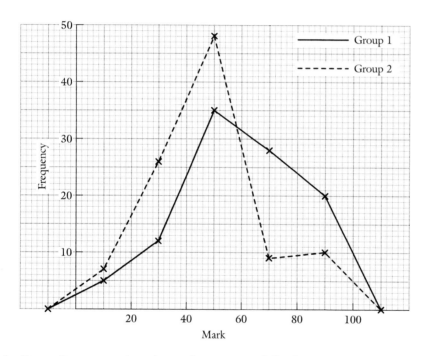

b Group 2 appear to have been given a more difficult examination paper than Group 1.

EXERCISE 8.7

1 Draw the frequency polygon for the data given below.

The speeds of 100 cars on a motorway were recorded.
The data found was:

Speed (mph)	30–40	40–50	50–60	60–70	70–80	80–90
Number of cars	2	11	35	42	9	1

2   A teacher noted the rates of absence from her Maths class on Mondays and Fridays. The results are given on the table below.

Number absent from class	0	1	2	3	4	5	6	7	8	9	10	11
Monday frequency	3	6	6	7	4	4	3	0	0	0	0	0
Friday frequency	0	2	2	3	4	5	2	0	6	3	2	1

a   Draw the frequency polygon for each day, using the same axes.

b   Comment on the absence rates for the two days.

3   The table below shows the number of letters per word in 100 words of two books.

No. of letters	1	2	3	4	5	6	7	8	9	10	11	12
Frequency, Book A	3	12	28	7	14	11	6	6	7	4	1	1
Frequency, Book B	6	8	37	22	7	11	6	0	1	2	0	0

a   Draw the frequency polygons using the same axes.

b   One extract was taken from a child's story and the other from an adult science fiction story.
State which is which, giving reasons for your decision.

4   The number of new designs being introduced by two furniture manufacturers was:

	1994	1995	1996	1997	1998	1999
Modern Design	8	7	8	5	2	1
Traditional Style	4	4	3	4	6	7

Draw the frequency polygons and comment on your results.

5   Choose two daily newspapers, one full size and the other a tabloid. Compare them by drawing frequency polygons of the number of words per sentence in one hundred sentences taken from similar sections in each newspaper.
(If you keep these results, they could be used in future work to calculate means, medians, and standard deviations.)

6   The table below shows the population of males and females in 1993 in the UK.

Age	0–4	5–15	16–44	45–59	60–64	65–79	80+
Male population (thousands)	1929	4276	12 258	5270	1355	3042	726
Female population (thousands)	1834	4059	11 840	5312	1418	3840	1643

(Source: *The Office of Population Censuses and Surveys*)
Draw the frequency of polygons and comment on your results.

7 A casino carried out an experiment using eight dice to find the number of sixes in each throw of the dice.

The objective was to see if the experimental results matched the expected theoretical calculations.

The frequency polygon shows the expected results.

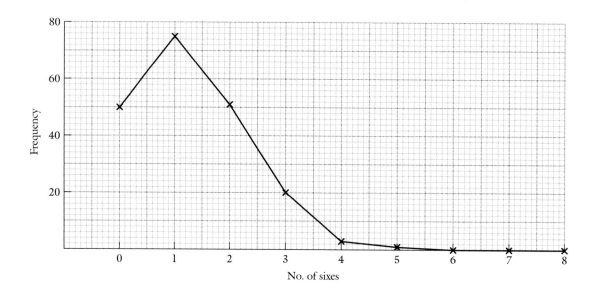

The experimental results were:

Number of sixes	0	1	2	3	4	5	6	7	8
Frequency	50	75	51	20	3	1	0	0	0

a Copy the given graph and plot the frequency polygon for the experimental results of the same axes.

b Comment on the two sets of results.

8 The number of faults on a car production line in two successive months, before and after a new model was introduced, was:

Number of faults	0	1	2	3	4	5	6	7	8	9	10
Frequency (old models)	0	0	1	4	5	8	8	2	1	0	2
Frequency (new model)	4	8	5	11	1	0	2	0	1	0	0

Draw the frequency polygons and comment on your results.

8.10   Stem-and-leaf diagrams

A stem-and-leaf diagram is another way of representing data by means of a diagram.

EXAMPLE

The number of letters in fifteen post boxes when the morning collection was made, was recorded as:

31, 49, 28, 33, 36, 41, 46, 36, 35, 27, 31, 35, 44, 25, 54

A stem-and-leaf diagram showing this data is:

Number of letters

KEY: 2 | 4 means 24

```
            5 | 4
Stem →      4 | 9 1 6 4      ← Leaves
            3 | 1 3 6 6 5 1 (5)
            2 | 8 7 5
```

The first post box had 31 letters in it. The first digit, 3, is the steam and the second digit, 1, is the leaf.

All the data is plotted in a similar way, so that the circled digit, 5, represents the number 35 (as the steam is 3 and the leaf is 5).

As the numbers in the leaves are not in numerical order, this is said to be an **unordered stem-and-leaf diagram**. When constructing such a diagram, it is common to redraw it with the leaves in numerical order. The above stem-and-leaf diagram would become

Number of letters

KEY: 2 | 4 means 24

```
5 | 4
4 | 1 4 6 9
3 | 1 1 3 5 5 6 6
2 | 5 7 8
```

This is known as an **ordered stem-and-leaf diagram**.

Notice that each stem-and-leaf diagram needs a heading and a key.

EXERCISE 8.8

1   The numbers of passengers on a small 72 seater aeroplane in its 20 short flights on one day were recorded as:

33, 37, 41, 45, 61, 72, 68, 72, 65, 61,

65, 39, 41, 48, 53, 54, 57, 49, 41, 31

Draw a stem-and-leaf diagram to show this data.

2   The numbers of supporters of a football club who travelled on Byron's coaches to away matches were:

> 83, 76, 92, 41, 62, 71, 59, 63, 70, 81,
> 74, 65, 67, 58, 49, 63, 71, 49, 54, 61

Draw a stem-and-leaf diagram to show this data.

3   The numbers of tourists travelling on a pleasure boat in the Solent on 14 days in August were:

> 71, 49, 82, 61, 91, 85, 89, 74, 92, 75, 64, 48, 52, 57

Draw a stem-and-leaf diagram to show this data.

4   The weight of packets, in kg, received by a small company were recorded as:

> 2.1, 4.8, 7.1, 5.2, 2.7, 3.5, 3.6, 4.1, 5.3, 4.7,
> 3.8, 2.7, 2.8, 4.7, 3.9, 6.2, 5.3, 6.8, 4.8

Using a key 4|3 to mean 4.3, draw a stem-and-leaf diagram to show this data.

Back-to-back stem-and-leaf diagram

A back-to-back stem-and-leaf diagram shows the information relating to two similar sets of data which could normally be shown in two separate stem-and-leaf diagrams.

EXAMPLE

The ages in years of holidaymakers in two small hotels on 1 August 2000 were:

Hotel Du Parc
> 37, 28, 41, 71, 62, 47, 54, 49, 63, 41, 38, 75, 51, 49,
> 47, 65, 58, 57, 51, 68, 72, 35, 51, 46, 52, 57, 64, 65

Le Soleil Hotel
> 52, 49, 75, 35, 36, 49, 25, 29, 31, 37, 28, 39, 62, 41, 25, 28, 32,
> 71, 36, 39, 29, 41, 31, 32, 35, 41, 59, 48, 72, 27, 41, 37, 25

Showing these ages in a stem-and-leaf diagram for Hotel Du Parc:

(The unordered diagram)
Ages of Holidaymakers
KEY: 2|4 means 24

7	1 5 2
6	2 3 5 8 4 5
5	4 1 8 7 1 1 2 7
4	1 7 9 1 9 7 6
3	7 8 5
2	8

(The ordered diagram)
Ages of Holidaymakers
KEY: 2|4 means 24

7	1 2 5
6	2 3 4 5 5 8
5	1 1 1 2 4 7 7 8
4	1 1 6 7 7 9 9
3	5 7 8
2	8

For Le Soleil Hotel the stem-and-leaf diagrams are:

(The unordered diagram)

Ages of Holidaymakers

KEY: 2 | 4 means 24

7	5 1 2
6	2
5	2 9
4	9 9 1 1 1 8 1
3	5 6 1 7 9 2 6 9 1 2 5 7
2	5 9 8 5 8 9 7 5

(The ordered diagram)

Ages of Holidaymakers

KEY: 2 | 4 means 24

7	1 2 5
6	2
5	2 9
4	1 1 1 1 8 9 9
3	1 1 2 2 5 5 6 6 7 7 9 9
2	5 5 5 7 8 8 9 9

A back-to-back steam-and-leaf diagram shows both of the ordered data on the same stem; Hotel Du Parc is shown on the left in order *away* from the stem and Le Soleil Hotel is shown on the right of the stem:

Ages of Holidaymakers

Hotel Du Parc   Hotel Le Soleil

KEY: 2 | 4 means 24

5 2 1	7	1 2 5
8 5 5 4 3 2	6	2
8 7 7 4 2 1 1 1	5	2 9
9 9 7 7 6 1 1	4	1 1 1 1 8 9 9
8 7 5	3	1 1 2 2 5 5 6 6 7 7 9 9
8	2	5 5 5 7 8 8 9 9

From this back-to-back stem-and-leaf diagram we can see at a glance that the guests at the Le Soleil Hotel tend to be younger than those in the Hotel Du Park.

EXERCISE 8.9

1   The numbers of cars passing a counter on a road in the New Forest per half hour period from 8 am until 8 pm on a particular Monday and Saturday in June were:

Monday	44	39	31	21	19	17	25	15	29	27	24	15	13	17	19	28	34	32	29	34	33	21	15	11
Saturday	3	4	11	31	42	47	51	49	56	51	67	71	49	53	51	69	51	42	41	39	28	31	29	25

Draw a back-to-back stem-and-leaf diagram to show this data.

2   The numbers of passengers in an aeroplane in the Caribbean on 15 short flights were recorded on a Wednesday and a Sunday as follows:

Wednesday	8	12	15	21	17	19	23	31	31	28	15	11	16	9	6
Saturday	19	28	27	29	28	30	30	31	29	24	21	25	17	15	12

a   Draw a back-to-back stem-and-leaf diagram to show this data.

b   What conclusions can you draw from this diagram?

3 During October the number of hours of sunshine recorded each day in two resorts, one in the Portugal and the other in the South of England, was:

Portugal	8.1	7.9	6.7	5.4	4.9	5.2	4.1	3.1	1.9	6.9	9.2	8.7	8.5	9.2	8.7	7.4
	5.2	4.1	1.1	4.2	1.5	7.2	8.4	7.9	8.3	9.2	9.1	8.9	8.8	7.1	8.2	
England	5.8	4.7	3.3	3.2	3.7	3.8	3.9	0.1	0.2	0.1	1.1	2.3	3.7	3.5	3.6	3.3
	4.1	8.1	7.1	3.9	45	4.7	3.9	4.8	2.1	1.7	1.5	1.9	4.9	2.8	2.9	

a Using a key 4 | 3 to mean 4.3, draw a back-to-back stem-and-leaf diagram to show this data.

b What conclusion can you draw from the diagram?

4 The number of male and female customers in "le Bistro" restaurant were recorded for a three-week period:

Male	21	15	17	19	35	42	51	34	47	25	31	25	17	21	29	38	52	47	8	38	28
Female	24	17	19	31	21	12	8	25	19	34	15	21	11	32	24	17	12	21	38	17	31

Draw a back-to-back stem-and-leaf diagram to show this data.

8.11   Shapes of distributions

Histograms and frequency polygons are also used to illustrate the shape of a frequency distribution.

Many frequency distributions have a recognisable shape which can easily be described.

1   A **symmetrical** distribution has the same shape either side of a central vertical line.
Example: IQ scores in the population.

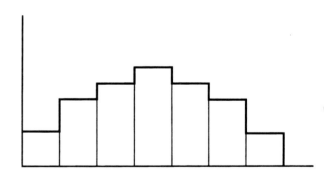

2   A **positively skewed** distribution has a 'tail' on the right of the graph.
Example: Results on a difficult exam paper.

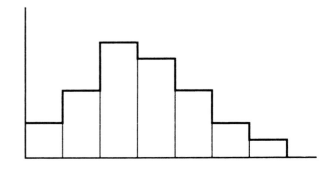

3 A **negatively skewed** distribution has a 'tail' on the left of the graph.
Example: Results on a relatively easy exam paper.

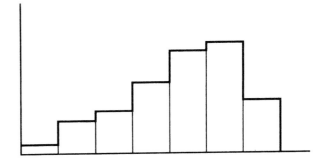

4 A **bimodal** distribution has two 'peaks'.
Example: Lengths of strides of female and male students.

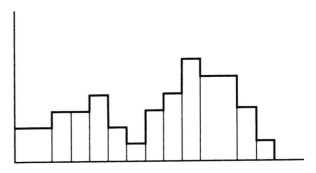

09 Averages and Range

Frequency distributions and graphs, such as dual bar charts and frequency polygons may be used to compare sets of data. It is also very useful to be able to compare a single, 'typical', statistic from one set of data with a single, 'typical', statistic from another set of data.

This statistic must be representative of the distribution. For this reason it is usually located at or near the centre of the distribution and is called a **measure of central location** or **average**.

The most commonly used averages are the **mean**, **mode** and **median**.

9.1 The arithmetic mean

The **arithmetic mean**, which is usually just referred to as the mean, is the most widely used average.

To calculate the mean, the total of the values is found and this is 'shared out' equally by dividing by the total number of values.

EXAMPLE

A company owns five nursing homes. The number of residents in the five homes is 45, 21, 41, 31 and 38.
What is the mean number of residents per home?

$$\text{Total number of residents} = 45 + 21 + 41 + 31 + 38$$
$$= 176$$
$$\text{Number of homes} = 5$$
$$\therefore \text{ Mean number of residents} = \frac{176}{5} = 35.2$$

This means that, if the five homes had the same number of residents, each home would cater for approximately 35.

Note. Although it is impossible to have 35.2 residents, it is usual to leave a mean as a decimal answer to show that it is a calculated value and to allow more accurate comparisons to be made.

The mean is often denoted by the symbol $\bar{x}$ and given in formula form as

$$\textbf{Mean} = \frac{\Sigma x}{n}$$

Σ is a capital Greek letter (sigma). $\sum x$ means the sum of all the terms and n is the number of terms.

EXERCISE 9.1

1   The weekly wages of ten workers were:

£110, £115, £135, £141, £119, £152, £144, £128, £117, £139.

Find the mean wage.

2   The wind speed (in mph) at 8 am on a particular day, was recorded at a number of measuring stations as:

88, 74, 61, 92, 48, 59, 71, 80, 70, 51, 48, 45, 75, 80, 82.

Find the mean wind speed.

3   A rugby team scores 37, 21, 64, 0, 18, 7, 35, 49, 28, 51, 82, 71 points in 12 successive matches. What is its mean score?

4   Eight people were asked their ages, and the replies were 37, 41, 29, 17, 15, 21, 32, 38. John claims that their average age is over 29. Why is he correct?

5   Five men have a mean height of 1.95 m.

Four women have a mean height of 1.72 m.

What is the mean height of the nine people?

6   The attendances at an art exhibition on six successive days were 130, 97, 110, 78, 64 and 150.

What was the mean daily attendance?

7   A quilter bought eight odd lengths of material from the ends of rolls to use for patchwork. She was charged for eight metres.

The actual length of the pieces were 1.50 m, 0.75 m, 1.20 m, 0.90 m, 1.30 m, 1.25 m, 1.55 m and 1.90 m.

a   What was the mean length of a piece of material?

b   Did she make a 'good buy'?

8   A secretary typed four documents in one hour. The numbers of words (to the nearest 10 words) were 1200, 850, 1570 and 880.

a   What was the average number of words per document?

b   What was the secretary's average typing speed in words per minute (w.p.m.)?

9   A canteen's food bills for one week were £498, £529, £384, £366, £620, £591, £485.

a   What was the mean cost per day?

The number of employees served on each day was 109, 110, 84, 88, 122, 14, 82.

b   What was the average cost per head?

10   In six successive weeks, the number of cases dealt with by a probation officer was 12, 15, 13, 11, 14, 16.

What was the mean number of cases per week?

11   Five overweight people volunteered to go on a sponsored diet. Their initial weights were 183 lb, 227 lb, 138 lb, 199 lb and 150 lb.

After three months, their weights were 141 lb, 180 lb, 126 lb, 146 lb and 150 lb.

a   What was the total weight loss?

b   What was the mean weight loss per person?

c   What was the mean weight loss per week? (Assume 1 month = 4 weeks)

12   The map shows the temperatures in various parts of Britain on a given day.

a   What was the mean temperature?

b   What was the mean temperature for:
   (i) the North?
   (ii) the South?

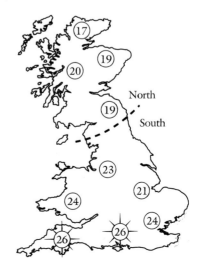

13 In an athletic match, in 1988, the winning distances for the discus were 52.78 m, 51.78 m, 50.77 m.

In 1998 the winning distances were 58.82 m, 57.48 m, 57.38 m.

a What was the mean distance in 1988?

b What was the mean distance in 1998?

c Comment on the results.

14 Twelve cars of the same model were fuelled with exactly one gallon of petrol. The cars were then driven over the same course at a steady speed of 56 m.p.h. until the petrol ran out. The number of miles travelled by each car was recorded:

38, 41, 44, 38, 35, 42, 41, 39, 44, 43, 46 and 40.

Is the manufacturer justified in claiming that the number of miles per gallon, at 56 m.p.h., is 41 for this model of car?

15 A time study was carried out at a factory to find the times taken for each process. Six fitters were timed assembling a piece of machinery. Their times (in minutes and seconds) were:

6:50, 5:35, 7:05, 5:10, 6:20 and 5:30.

What was the mean time for the job?

9.2   The mode

The **mode** is the number which occurs most frequently. Suppose seven students scored as follows in a test:

$$2, 3, 6, 7, 7, 8, 9$$

The mean score here is $\dfrac{42}{7} = 6$.

The **mode**, the number which occurs most frequently, is 7.

Some distributions can have more than one mode. For example, the numbers of people hiring a firm's mini bus were:

3, 3, 4, 5, 7, 8, 8, 8, 10, 11, 13, 13, 14, 15, 15, 15.

Both 8 and 15 are modes.

This distribution is said to be **bimodal**.

EXAMPLE 1

The number of days spent in hospital by patients on a surgical ward were

2, 3, 3, 2, 4, 4, 3, 5, 10, 3, 2, 4, 3, 4.

Find the mode.

When the numbers of days are written in ascending order:

2, 2, 2, 3, 3, 3, 3, 3, 4, 4, 4, 4, 5, 10

it can be seen that the number occurring most often is 3.
This means that the mode or modal number of days is 3.

Modal class

When the values are grouped the mode is replaced by a **modal class**, which is the group of values occurring most frequently.

EXAMPLE 2

What is the modal class in the following table, which shows the time spent by customers in a leisure centre.

Time spent (minutes)	Frequency
Less than 15	6
15 –	10
30 –	13
45 –	12
60 –	16
75 –	20
90 –	21
105 –	17
120 – 135	14

There are 21 people who spent between 90 and 105 minutes in the leisure centre.
This is the highest frequency.
∴ The class 90 – 105 is the modal class.

9.3   The median

The **median** is the value of the 'middle' observation when the observations are placed in numerical order.

To calculate the median, list all the observations given in numerical order (usually ascending), and the median is the value of the middle one.

A median cannot, therefore, be found for qualitative data.

EXAMPLE 1

The cost of diesel fuel (per litre) at five garages is 76.8p, 74.2p, 82.9p, 83.7p and 78.9p.
What is the median cost?

Write the costs in ascending order:

$$74.2 \quad 76.8 \quad 78.9 \quad 82.9 \quad 83.7$$

Find the cost in the middle position:

$$74.2 \quad 76.8 \quad \boxed{78.9} \quad 82.9 \quad 83.7$$

∴ The median cost is 78.9p per litre.

EXAMPLE 2

A sixth garage charges 76.9p per litre for diesel. The costs are now 76.8p, 74.2p, 82.9p, 83.7p, 78.9p and 76.9p.
What is the median cost?

Because there is an even number of costs, there are *two* costs in the 'middle':

74.2   76.8   (76.9   78.9)   82.9   83.7

The convention is to average these two 'middle' values.

The average (mean) of 76.9 and 78.9 $= \dfrac{76.9 + 78.9}{2} = 77.9$

$\therefore$ The median cost is 77.9p per litre.

EXERCISE 9.2

1   A die was thrown 12 times, and the scores were 2, 4, 1, 3, 4, 1, 5, 6, 6, 4, 2, 5.

   a What was the modal score?

   b What was the median score?

2   Eleven cars have the following colours: blue, black, red, white, silver, blue, red, grey, black, blue, green.

   What is the modal colour?

3   The IQs of 70 students were recorded, to the nearest integer, as:

IQ	Frequency
95 – 99	5
100 – 104	7
105 – 109	18
110 – 114	21
115 – 119	7
120 – 124	4
125 – 129	5
130 – 134	1
135 – 140	2

What is the modal class?

4   The number of cars per hour on a country road during the hours of daylight were recorded as 11, 13, 15, 11, 17, 12, 18, 14, 7, 9, 14, 16, 7, 11.

   a Find the mean number of cars.

   b Find the modal number of cars.

5   The weights of parcels (in kg) delivered to a library were:

7.4, 8.2, 11.1, 7.8, 2.5, 5.6, 7.1, 8.9, 2.3, 2.7, 2.9, 4.1.

Find the median weight.

6   What was the modal number of artists named by the general public in question 1 of Exercise 7.1 (page 37)?

7   A photographer arranges a family group with the tallest members of the family in the centre and the smallest on the outside:

For his next photograph, he arranges the family in ascending order of height.

   a Who are now in the centre of the line?

   b What is the median height of the family?

   c Which average is not a representative value?

8 The price of a loaf of bread (in pence) at 10 shops was found to be:

44, 49, 51, 68, 62, 44, 69, 51, 44, 47.

  a What was the median price?

  b What was the modal price?

9 Five managerial staff have car allowances of £35 000, £23 000, £21 000, £18 000, £28 000. Six office staff each have a car allowance of £10 000.

  a What is the median allowance?

  b What is the modal allowance?

  c What is the mean allowance?

  d Which average best represents the allowances?

10 What was the modal class in:

  a Question 5   **b** Question 6

  of Exercise 7.1 (page 37)?

11 The ages at which eleven patients were first diagnosed as diabetic were:

5, 12, 16, 18, 25, 16, 17, 57, 32, 60, 61

  a What was the median age?

  b What was the modal age?

12 In question 7 of Exercise 7.1 (page 38), what was the modal holiday destination for a wedding?

13 The cost of a ski lift pass in seven popular ski resorts is:

£112, £160, £134, £100, £88, £112, £120

  a What is the modal cost?

  b What is the median cost?

14 The number of missing parts in packs of kitchen units was recorded for a sample of 10 packs as:

4, 3, 6, 4, 1, 0, 1, 2, 0, 4

  a Find the modal number of missing parts.

  b Find the median number of missing parts.

15 The time taken (in minutes and seconds) to carry out the test on packs of kitchen units was recorded for six quality controllers as:

5:20, 6:40, 5:30, 4:24, 3:58, 4:30

Find the median time taken.

9.4 The use of mean, mode and median

The three averages are useful in different contexts.

- If you were an employer considering the production capacity of your works, it would be helpful to use the *mean* of past production, as this would give you a good idea of the number of goods you can produce.

- If you were a shopkeeper wanting to keep a minimum stock of shirts to sell, the *mode* would be the best to use, as this will tell you which shirts you are most likely to sell.

- If you were a union wage negotiator, the *median* salary would be appropriate to use, because the few high wage earners would not then affect your 'average' of the wages paid.

EXERCISE 9.3

In each of the following situations, decide which of the mode, the median and the mean would be the most appropriate to use. (You are not required to find any values.)

1 A group of artists working on a sculpture each suggested how it should be positioned.

A typical value is required.

2 Students in an art class were asked how long they had spent on their last project.

The replies (in hours) were:
3, 4, 3, 5, 10, 4, 4, 3

3   Witnesses to a shoplifting incident were asked how many thieves were involved.

Their answers were: 1, 2, 2, 3, 3, 3, 3, 4

4   A student tries to decide if he is fairly paid for his part-time job.

He asks some friends how much per hour they are paid for similar work.

5   The electricity consumption for a dental practice is noted for the months from October to March.

A typical value is required.

6   Witnesses to a heart attack were asked to estimate how long the sufferer was unconscious. The answers (in minutes) were:

8   $8\frac{1}{2}$ $7\frac{1}{2}$ 9 $9\frac{1}{2}$

The time needs to be determined as accurately as possible.

7   Ten students were asked how much they spent on their last year's holiday.

The answers were:
£250, £162, £340, £140, £860, £300, £98, £380, £970 and £105

8   Caravaners returning from Europe on a cross-channel ferry were asked what distance they had travelled during their holiday.

A typical value was required.

9   A manager and his deputy managers were asked to estimate the cost of installing a new production process.

The average of the estimates is to be taken as a working value.

10 Matches are packed into boxes on which is written 'average contents 280 matches'.

9.5   The mean and median of a frequency distribution

In this example the data is recorded in a frequency table:

EXAMPLE 1

Find the mean and mode of the following scores:

Score	Frequency
1	3
2	5
3	11
4	1
5	5

The score of 2, for example, occurred five times, but instead of totalling $2 + 2 + 2 + 2 + 2$, it is quicker to multiply 2 by 5. Similarly, instead of totalling $3 + 3 + 3 + \ldots$ eleven times, it is quicker to calculate 3×11.

The table can be extended like this:

Score x	Frequency f	Score × Frequency xf
1	3	3
2	5	10
3	11	33
4	1	4
5	5	25
Totals:	$\Sigma f = 25$ (Total frequency)	$\Sigma xf = 75$ (Total of 25 scores)

Mean score $= \dfrac{\sum xf}{\sum f}\left(\text{i.e. } \dfrac{\text{Total of 25 scores}}{\text{Total frequency}}\right) = \dfrac{75}{25} = 3$

The highest frequency is 11.
∴ The mode is 3.

EXAMPLE 2

Thirty households were surveyed, and the number of children in each household was recorded. Find the mean and the mode.

No. of children in each family		0	1	2	3	4	5
Frequency			4	6	13	4	2	1

No. of children in each family x	Frequency f	xf
0	4	0
1	6	6
2	13	26
3	4	12
4	2	8
5	1	5
Totals:	30	57

$$\text{Mean} = \frac{\Sigma xf}{\Sigma f}$$

$$= \frac{57}{30}$$

$$= 1.9 \text{ children}$$

There were 30 households. Therefore the median number of children is between the numbers in the 15th and 16th households. The cumulative frequencies (see Unit 10) are 4, 10, 23, ... , i.e., both the 15th and 16th households contained 2 children.

∴ The median is 2 children.

Calculating the mean of a grouped distribution is more complicated. How to deal with the values when they are in classes is shown in the following example.

EXAMPLE 3

Thirty bushes were measured. Their heights (in cm) were grouped as shown.
Find the mean height and the modal class.

Height	5–15	15–25	25–35	35–45	45–55
Frequency	6	4	15	3	2

Since there is a spread in each group, it is impossible to determine the exact mean. We can only find an approximation to the mean.

It is assumed that each bush has a height equal to the middle of the range in which it lies; for instance, the 4 bushes with height 15–25 cm are all assumed to have height 20 cm, which is the **mid-interval** of the range 15–25 cm.

Length (cm)	Mid-interval (cm) x	Frequency f	Mid-interval × Frequency xf
5–15	10	6	60
15–25	20	4	80
25–35	30	15	450
35–45	40	3	120
45–55	50	2	100
Totals:		30	810

$$\text{Mean} = \frac{\text{Total of `mid-interval} \times \text{Frequency'}}{\text{Total frequency}} = \frac{\Sigma xf}{\Sigma f} = \frac{810}{30} = 27\,\text{cm}$$

The modal class is 25–35 cm. (There are 15 bushes in this class, more than in any other class.)

EXAMPLE 4

In one hour, twenty planes arriving at Heathrow were early. The number of minutes early (to the nearest minute) were grouped as shown. Find the mean number of minutes early.

Time (min)	0–3	4–8	9–13	14–18
Frequency	4	7	8	1

As in Example 3 there is a spread in each group. The lower and upper class boundaries need to be found carefully. The group 0–3 (minutes) will range from 0 minutes (early) to 3.5 minutes (early). Therefore, the four planes in this group are all assumed to be 1.75 minutes early, which is the mid-interval of the range 0–3.5.

The group 4–8 ranges from 3.5 to 8.5 minutes early. Hence, these seven planes are all assumed to be 6 minutes early. (6 is the mid-interval of 3.5 and 8.5.)

No. of minutes early	Mid-interval (in minutes) x	Frequency f	Mid-interval × Frequency xf
0–3	1.75	4	7
4–8	6	7	42
9–13	11	8	88
14–18	16	1	16
Totals		20	153

$$\therefore \quad \text{Mean} = \frac{\Sigma xf}{\Sigma f}$$
$$= \frac{153}{20}$$
$$= 7.65\,\text{minutes}$$

Note. In Examples 3 and 4 the distributions are continuous.

No problems are caused in Example 3 by including the height 15 cm in the group 5–15 and in the group 15–25 for heights of bushes, since the probability of finding a bush exactly 15.000...m high is nil.

When a distribution is discrete, you should *not* use 5–15 and 15–25, etc., since if an item is exactly 15, it is not clear into which group it should be placed.

If you were considering the number of people on a bus, you could use 5–14, 15–24, 25–34, etc. The range 5–14 would have a mid-interval of 9.5, which would be used to find the mean.

EXERCISE 9.4

1 An agricultural researcher counted the numbers of peas in a pod in a certain strain as follows:

No. of peas	3	4	5	6	7	8
No. of pods	5	5	20	35	25	10

Find the mean number of peas per pod.

2 The Ace Bus Company went through a bad patch when its buses always left the city centre late. This grouped distribution table shows how late:

Minutes late	0–10	10–20	20–30	30–40	40–60
Frequency	5	8	21	14	5

Find the mean number of minutes late.

3 The numbers of words per sentence on a page of a book were:

No. of words	1–3	4–6	7–9	10–12	13–15
Frequency	3	38	59	27	4

Find the mean length of a sentence.

4 The numbers of faults found by a potter in glazed pots after being fired in the kiln were:

No. of faults	0	1	2	3	4	5
Frequency	15	8	7	6	3	1

a What was the mean number of faults?
b What was the median number of faults?
c What was the modal number of faults?
d Which average best represents the data?

5 The number of weddings per week attended by a photographer was:

No. of weddings	0	1	2	3	4	5	6	7	8
No. of weeks	2	8	3	5	10	7	9	5	3

a What was the mean number of weddings attended per week?
b What was the median number of weddings attended per week?

6 The numbers of times the photocopier was used on one day by the staff in an office were recorded:

No. of times used	2	5	6	7	8	9	10
Frequency	1	3	3	2	4	4	1

Find **a** the median, **b** the mean number of times the photocopier was used on this day.

7 The manager of an electrical business recorded the number of TV sets brought in for repair each week day over a three-month period. The results of the survey were:

No. of TV sets	0	1	2	3	4	5	6
Frequency	5	7	12	15	19	16	4

a What was the median number of TV sets?
b What was the mean number of TV sets?

8 The numbers of children per family on a housing estate were recorded as follows:

No. of children	0	1	2	3	4
No. of families	12	15	5	2	1

Find **a** the mean, **b** the median number of children per family.

9 At a research centre for the common cold, 25 volunteers were exposed to the cold virus under controlled conditions. The time taken (in days) for the first symptoms of a cold to appear were (one person did not catch a cold):

No. of days	2	3	4	5	6
Frequency	2	5	8	7	2

a Find the mean number of days.
b Find the median number of days.
c If the person who did not catch cold were included in the data, which average would not be changed?

10 The goals scored in 42 football league matches on Saturday 26 March 2000 were:

No. of goals per match	0	1	2	3	4	5	6
No. of matches		1	9	10	12	6	3	1

a Find the mean number of goals per match.

b Find the median number of goals per match.

11 The numbers of injuries per week sustained at a dry-ski school were:

No. of injuries	0	1	2	3	4	5	6	7
No. of weeks	3	10	9	13	5	7	4	1

a Find the mean number of injuries per week.

b Find the median number of injuries per week.

12 Find **a** the mean, **b** the median number of faults in the micro chips (see question 10, Exercise 7.1, page 38).

No. of faults	0	1	2	3
Frequency	31	13	5	1

c Which average should be used to compare this data with a similar set?

13 During a survey into the ages of employees in a factory, the ages of the apprentice tool makers were recorded as:

Age (years)	17	18	19	20	21	22	23
Frequency	4	4	5	2	3	2	1

a Find the median age.

b Find the mean age.

c Comment on your results.

9.6 Range

Averages are used to represent sets of data or to compare them, but an average on its own does not give sufficient information about the distributions.

Suppose we are comparing the climate of two places in Turkey. Town A has an average yearly temperature of about 12°C and town Z has an average yearly temperature of about 13°C. From this we might suppose that the climates are similar and that town Z possibly lies south of town A.

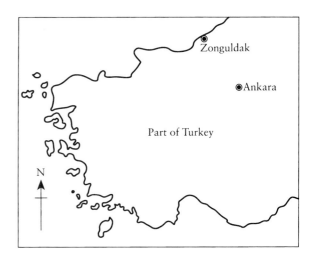

Part of Turkey

In fact, town Z is Zonguldak, which is on the Black Sea coast, and town A is Ankara, which is further south in the interior of Turkey.

The monthly temperature distributions for the two towns are:

Month	J	F	M	A	M	J	J	A	S	O	N	D
Temperature in Ankara (°C)	−0.2	1.2	4.9	11.0	16.1	20.0	23.3	23.4	18.4	12.9	7.3	2.1
Temperature in Zonguldak (°C)	6.0	6.3	7.0	10.5	15.0	19.2	21.7	21.6	18.4	15.0	11.5	8.5

We can see that the climate in Ankara is more variable than in Zonguldak. There is a larger difference between summer and winter temperatures, whereas in Zonguldak there is a smaller spread of temperatures and a more equable climate.

We need a statistic which will measure this spread of values. The term we use to describe spread is **dispersion**, and there are several ways of measuring it.

The simplest measure of spread is the **range**, which is the difference between the lowest and the highest values.

The range of temperatures for each town is

Ankara: Range $= 23.4 - (-0.2) = 23.6°C$
Zonguldak: Range $= 21.7 - 6.0 = 15.7°C$

There is a problem with using the range because it uses only two values and so it can be distorted by a very high of low value.

EXERCISE 9.5

1   A fashion designer makes a particular style of skirt in various lengths:
    16", 19", 24", 28", 32", 34" and 36".
    What is the range of lengths?

2   Five bank clerks took a 'tea break' at 10.30 am. They returned to their desks at the following times:
    1038, 1040, 1042, 1045 and 1048
    a Find the time taken for each 'tea break' and calculate the mean.
    b Find the range of the times.

3   Eight patients suffering from lung cancer were asked how many cigarettes per day they smoked. The replies were:
    40, 44, 25, 0, 50, 30, 35, 30.
    Calculate (i) including the non-smoker, (ii) excluding the non-smoker:
    a the mean number of cigarettes smoked per day.
    b the range of the number of cigarettes smoked per day.

4   The numbers of cars on a hovercraft in one day were:
    37, 28, 8, 5, 17, 39, 22, 10.
    Find the range.

5   The number of cars produced per year per employee by nine manufacturers in England were:
    7.1, 12.1, 8.1, 4.7, 11.6, 9.4, 3.6, 12.5, 3.9.
    Find the range.

6   Two designers collect shells, fossils, small stones, etc., with which to make jewellery and to decorate their designs.
    They collect from two beaches early in the morning. The numbers of items collected each day for a week are recorded below:

Day	Mon	Tue	Wed	Thu	Fri	Sat	Sun
Beach 1	20	27	49	71	62	24	15
Beach 2	43	39	29	46	51	37	42

    a Calculate the mean of each distribution.
    b Calculate the range of each distribution.
    c Which beach is the best for beach combing and why?

7   Two book clubs offer 'mystery parcels' of books for £8.50, stating that the minimum value of the contents is £15.

A survey of eight such parcels from each of the two clubs found that the actual value of the contents was:

Club 1 Value (£)	16.00	15.80	16.85	15.80	17.85	15.30	15.75	15.45
Club 2 Value (£)	15.95	16.90	16.85	17.85	15.25	17.50	17.00	17.25

a Calculate the mean and range of the data and comment on your results.

b Find the median value of the contents for each club.

c By comparing the mean and median, describe the distribution of each set of data.

8   A patient's blood pressures were recorded on 10 successive days.

Systolic blood pressure	118	126	119	125	127	126	120	136	120	106
Diastolic blood pressure	78	80	80	82	88	87	87	96	81	66

a Calculate the mean systolic and diastolic blood pressures.

b Find the range of:
(i) the systolic blood pressures   (ii) the diastolic blood pressures.
Normal levels are 110–140 for systolic and 70–90 for diastolic pressure.

c What can you conclude about this patient's blood pressure?

9   Two archers each shot six arrows at similar targets. The distance, in centimetres, of each shot from the centre of the target was measured and recorded:

Shot	1	2	3	4	5	6
1st archer	89	53	45	54	56	38
2nd archer	120	112	10	26	59	6

a Calculate the mean distance from the centre for each archer.

b Calculate the range for each archer.

c Which archer was the better shot?

10   The diameters (in millimetres) of ball bearings should be 50 mm.
Two samples of eight bearings, produced by two of the machines in question 11, Exercise 7.1 (page 38), had diameters as given below:

Machine 3	51.15	48.16	50.32	51.15	49.33	50.05	52.63	49.63
Machine 5	49.72	49.43	46.78	48.88	48.20	49.46	49.19	48.21

a Calculate the mean diameter for each machine.

b Calculate the range for each machine.

c Comment on the performances of the two machines.

10 Cumulative Frequency

The cumulative frequency is the total frequency up to a particular class boundary. It is a 'running total', and the cumulative frequency is found by adding each frequency to the sum of the previous ones.

10.1 The cumulative frequency curve (or ogive)

Virtually all cumulative frequency curves (or ogives) have an 'S' shape. How an ogive is built up will be seen in the following example.

EXAMPLE

The marks obtained by 100 students in an examination were as shown opposite.

Draw the cumulative frequency curve for this data.

Marks	No. of students (frequency)
0–10	1
11–20	2
21–30	13
31–40	24
41–50	32
51–60	16
61–70	11
71–80	1

First, we construct a new table. This keeps a running total of the frequencies in the second column.

From the cumulative frequency column we can see that one student has 10 marks or less, three students have 20 marks or less, sixteen students have 30 marks or less, and so on.

Marks	Cumulative (frequency)
0–10	1
0–20	1 + 2 = 3
0–30	3 + 13 = 16
0–40	16 + 24 = 40
0–50	40 + 32 = 72
0–60	72 + 16 = 88
0–70	88 + 11 = 99
0–80	99 + 1 = 100

Next we plot the cumulative frequencies against the **upper class boundaries**. For example, we plot 3 (students) against 20 (marks) and 16 (students) against 30 (marks). The completed graph then looks like this:

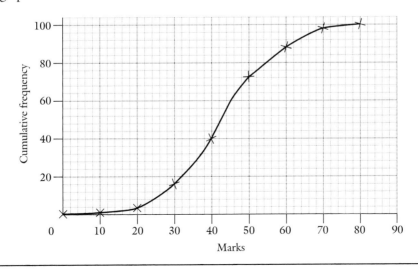

10.2   The median

The median mark is the mark obtained by the middle student. For example, if there are 100 students, the median student is the 50th (half-way point). (The strict definition would of course be the $50\frac{1}{2}$th student, but this accuracy cannot be obtained from a graph, and it is unnecessary at this stage.)

Suppose we wanted to find out the median mark in the example from section 10.1. We would look across from the 50 on the cumulative frequency axis to the curve, then read down vertically to the number of marks. In this case the median is 43 marks. This is illustrated below:

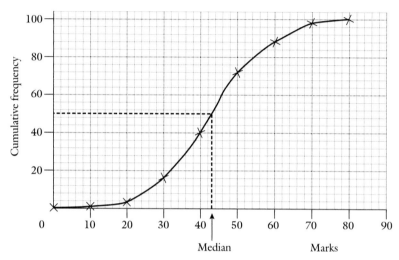

10.3   The interquartile range

(i) The **lower quartile** is the mark obtained by the student $\frac{1}{4}$ of the way along the distribution.
    There are 100 students in the example on page 85, so the 25th student is $\frac{1}{4}$ of the way up the cumulative frequency axis. From the graph, the 25th student has 35 marks, so the lower quartile is 35.

(ii) The **upper quartile** is the mark obtained by the student $\frac{3}{4}$ of the way up the cumulative frequency axis. This is the 75th student, whose mark is 52, so the upper quartile is 52.
    The upper and lower quartiles are illustrated below:

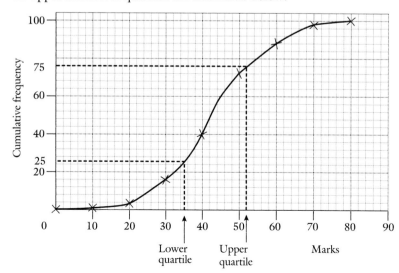

(iii) The **interquartile range** is the difference between the upper quartile and the lower quartile.

In our example, the interquartile range is:

$$52 - 35 = 17 \text{ marks.}$$

The **semi-interquartile range** is half the interquartile range, 8.5 in our example.

Note. Half the students have a mark between the lower and upper quartiles. Hence the interquartile range of 17 shows that half the marks obtained lie in an interval of 17. The semi-interquartile range of 8.5 shows that half the population lie (roughly) within 8.5 marks of the median mark.

10.4 Percentiles

We can use the example on page 85 to define a **percentile**. As the name suggests, percentiles divide the cumulative frequency distribution into 100 parts, just as the quartiles divide it into quarters. The 70th percentile, for example, is the highest mark obtained by 70% of the entry. From the cumulative frequency curve this mark is 49, so the 70th percentile is 49 marks (as shown below).

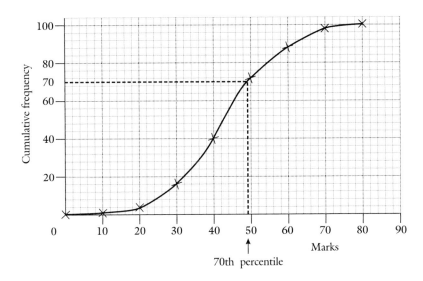

EXERCISE 10.1

In questions 1–4, calculate the cumulative frequencies, and draw the cumulative frequency curve. Hence find:

a the median   **b** the lower quartile   **c** the upper quartile   **d** the interquartile range.

Part **e** is given separately for each question.

1   The times taken for students to complete two questions were recorded as:

Time (in minutes)	0–20	20–25	25–30	30–35	35–40	40–45	45–50
No. of students	3	12	25	49	21	7	3

e Estimate the number of students who took more than 41 minutes.

2   The waist measurements of 100 people were recorded. The data found
were:

Waist (cm)	40–50	50–60	60–70	70–80	80–90	90–100
Frequency	2	11	25	41	20	1

    e (i) Estimate the number of people with waist measurement less than
53 cm.

       (ii) Estimate the number of people with waist measurement more
than 72 cm.

3   The annual gross wages of employees at a factory were:

Wage (£)	0–	8000–	10 000–	12 000–	14 000–	16 000–18 000
Frequency	1	15	34	58	27	3

    e Estimate the 80th percentile.

4   The marks gained by 90 students on a test (out of 100) were:

Marks	0–20	21–40	41–60	61–80	81–100
Frequency	2	15	32	33	8

    e (i) What percentage of students passed the exam if the lowest pass
mark was 37?

       (ii) Six students are given a distinction.
What was the lowest mark to obtain a distinction?

      (iii) Find the 90th percentile.

5   Potatoes are supplied to a greengrocer's shop in 50 kg bags. Each of the
potatoes in one of these bags was weighed to the nearest gram, and the
following table was drawn up:

Mass (g)	50–99	100–149	150–199	200–249	250–299	300–349
No. of potatoes	5	53	87	73	33	3

    a (i) Complete the cumulative frequency for these potatoes.

       (ii) Draw the cumulative frequency graph.

    b From your graph, estimate:

       (i) the median mass

      (ii) the number of potatoes weighing at least 225 g each.

    c For a party, Harry requires 50 baking potatoes, each weighing at least
225 g.
The greengrocer sells the potatoes without special selection.

       (i) Use your answer to **b**(ii) to estimate how many kilograms Harry
will have to buy.

      (ii) Taking the mean mass of the 50 baking potatoes to be 260 g,
estimate how many kilograms of potatoes he will have left over.

6   The time for 160 adults to complete an exercise schedule is shown on the frequency diagram.

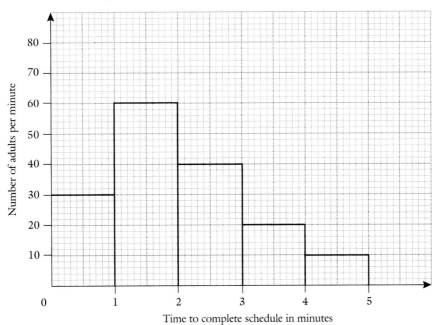

Time to complete schedule in minutes

a   Construct a cumulative frequency table for the data.

Time in minutes	Cumulative frequency
≤ 1	
≤ 2	
≤ 3	
≤ 4	
≤ 5	

b   Copy these axes and draw the cumulative frequency curve.

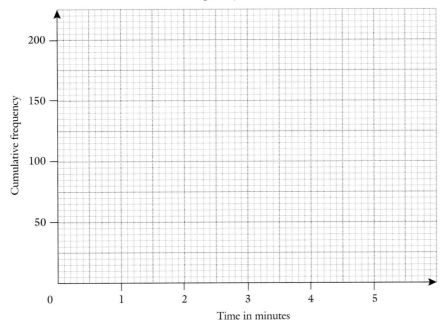

Time in minutes

c   Use your graph to estimate the time it would take 40% of the adults to complete the exercises.

d   Use your graph to find the number of adults who took **more** than 3.5 minutes to complete the exercises.

(SEG W95)

10.5 Box-and-whisker diagrams

A box-and-whisker diagram, sometimes known as a box plot, shows:

> (i) the median,
>
> (ii) upper and lower quartiles

and (iii) the maximum and minimum values of a distribution.

For the distribution in the example on page 85:

the median is	43
the upper quartile is	52
the lower quartile is	35
the maximum value is	80
the minimum value is	0

With a normal linear scale, from 0 to 80, draw a thin box from the lower quartile to the upper quartile as shown in the diagram alongside.

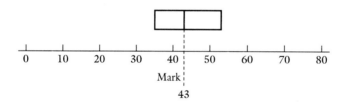

Insert a vertical line in the box at the median value.

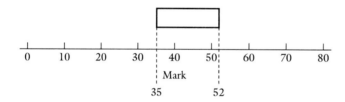

Now insert two lines or whiskers from the centre of each end of the box to the maximum and minimum values.

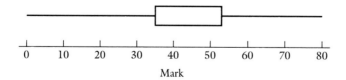

We have now constructed the box-and-whisker diagram relating to the distribution on page 85.

If the median is to the *left* of the centre of the box, the distribution is said to be *positively* skewed.

If the median is to the *right* of the centre of the box, the distribution is said to be *negatively* skewed.

EXAMPLE

Sixty trees in a plantation are measured.

Find

a the median,

b the interquartile range

c the range of the height of the trees which is shown in the diagram below.

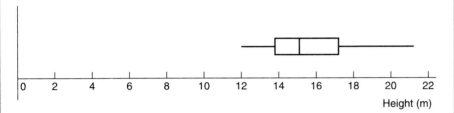

From the diagram:

a The median is 15.1 m.

b The upper quartile is 17.2 m.
   The lower quartile is 13.8 m.
   The interquartile range is 17.2 − 13.8 m = 3.4 m.

c The maximum height is 21.3 m.
   The minimum height is 11.9 m.
   The range is 21.3 − 11.9 m = 9.4 m.

Note. The box-and-whisker diagram may be drawn vertically instead of horizontally.

EXERCISE 10.2

1–5 Draw a box-and-whisker diagram for the data given for each of the Questions 1–5 in Exercise 10.1.

6 Draw a box-and-whisker diagram for the data given in Question 3 in Exercise 8.5.

7 The speeds in miles per hour of 32 cars on a motorway were recorded as:

68, 69, 74, 81, 61, 70, 65, 91, 58, 71, 68, 74, 82, 83, 55, 97, 82, 71, 68, 67, 75, 65, 61, 78, 71, 61, 62, 69, 51, 92, 41, 54

Draw a box-and-whisker diagram to show this data.

8 The numbers of passengers in a 72-seater aeroplane was noted on each of its 20 flights in one day. The numbers were:

48, 54, 72, 21, 37, 45, 72, 41, 59, 63, 54, 59, 68, 72, 70, 64, 41, 32, 24, 31

Draw a box-and-whisker diagram to show this data.

11 Correlation

All the graphs and charts used so far to represent data have been concerned with only one variable, e.g. height, holiday destination, number of goals scored. In many statistical surveys, however, the data collected connect two variables, e.g. height *and* weight; number of police cameras on a stretch of motorway *and* number of speeding convictions; age *and* reaction time.

Data of this type is collected because it is believed that there is a link between the two variables.

11.1   Scatter diagrams

The usual method of representing these types of data is by a **scatter diagram,** also called a **scatter graph** or **scattergram**.

EXAMPLE

A survey was carried out by a group of eight students in which the height and weight of each student was measured. The results were recorded in pairs (e.g. the student with height 164 cm weighed 58.2 kg).

Height (cm)	164	152	173	158	177	173	179	168
Weight (kg)	58.2	50.8	60.3	56.0	76.2	64.2	68.8	60.5

Display this data on a scatter diagram.

Two axes are drawn, one for the heights and one for the weights.
(It does not really matter which is which, but, as a general rule, the first set of data is recorded along the horizontal axis and the second set along the vertical axis.)

Each point is plotted using the paired data as the coordinates, i.e. for the student with height 164 cm and weight 58.2 kg, the coordinates are (164, 58.2).

The scatter diagram shows the height/weight of eight students.

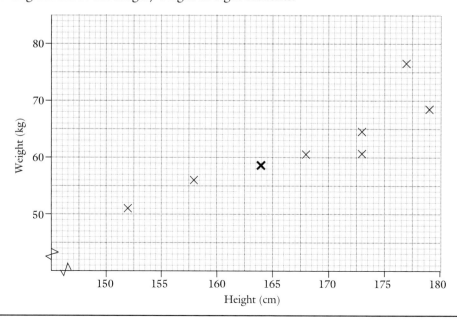

1   Draw scatter diagrams for the following data:

a
Mark on paper 1	15	20	14	5	24	10	19	26	30
Mark on paper 2	22	34	50	20	66	32	50	34	60

b
Time (minutes)	10	15	20	25	30	35	40
Weight of melting	3.9	3.5	2.7	1.5	1.2	1.0	0.8

c
Weight of parcel (kg)	1.0	4.0	5.5	3.5	5.0	5.5
Length of parcel (cm)	6	15	7	35	10	25
Weight of parcel (kg)	4.4	2.2	2.0	6.0		
Length of parcel (cm)	26	20	30	33		

2   Draw a scatter graph for the following data.

Shoe size	$4\frac{1}{2}$	10	5	9	7	6	$5\frac{1}{2}$
Handspan (cm)	20.2	21.6	17.3	19.6	21.2	21.2	19.4
Shoe size	10	8	9	5	5	11	
Handspan (cm)	22.0	19.5	23.7	19.5	20.2	23.0	

3   The engine capacities, in cubic centimetres (cc), and the corresponding acceleration times, in seconds, for several models of car in the range of one manufacturer are given below:

Engine size (cc)	900	1000	1100	1200	1350	1500	1600
Time (s)	20.0	15.9	13.9	13.3	11.4	12.4	10.7
Engine size (cc)	1750	1800	1900	2000	1100	1350	2000
Time (s)	12.3	7.0	8.1	6.8	14.1	9.4	9.9

Draw a graph for this data.

4   a   The data given below shows the percentage of breath-tests on motorists during the Christmas/New Year period which proved positive and the number of car accidents which involved injury during the same period. The figures given are for eleven counties in England and Wales.

% of positive tests	9.1	4.1	8.9	9.2	5.5	9.2	7.8
Injury accidents	65	49	46	97	25	96	56
% of positive tests	4.2	5.6	4.9	10.5			
Injury accidents	43	56	42	89			

Draw a scatter graph for these eleven points.

b   The same variables for another four counties are given below.

% positive tests	9.0	7.6	1.7	17.5
Injury accidents	5	112	63	43

Draw a scatter diagram for all 15 points and compare the two graphs.

11.2 Correlation

As might be expected, the scatter diagram in the example on page 92 shows that there is some link between height and weight but that it is not a very close one. For example, two people who are the same height may have very different weights. In general, however, the taller you are, the more you will weigh.

There are three basic types of scatter diagram:

Diagram (i)

The points are widely scattered. This shows that there is no link between the variables. For example, having a small head does not mean you are less intelligent than someone with a large head.

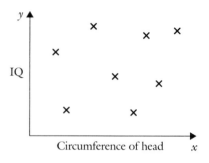

Diagram (ii)

This diagram might represent the number of police cameras on a motorway against the number of convictions for speeding. The points lie quite close together and show an upward trend. We might conclude that there is a weak connection, i.e. that the more cameras there are, the greater the number of motorists caught speeding.

However, the purpose of the cameras is not so much to catch motorists speeding as to deter them from speeding.

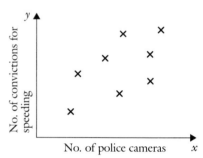

Diagram (iii)

This could represent the same variables after the cameras have been in place for a while and motorists are aware of them. There is still a link, but now the trend is downward, and we might conclude that the more cameras that are known to be operating, the less likely motorists are to drive fast and be caught speeding.

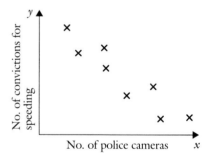

The closer a set of points lies to a straight line, the stronger the relationship between the variables, or, in statistical terms, the higher the **degree of correlation**.

Diagram (i) indicates that there is no correlation
Diagram (ii) indicates quite a weak positive correlation
Diagram (iii) indicates quite a high negative correlation.

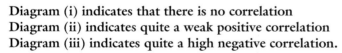
EXERCISE 11.2

1–4 For each of the questions in Exercise 11.1, state the type of correlation (none, positive, negative) and give some indication of the degree of correlation.

5 For each of the following sets of data, state whether there is no correlation, positive correlation or negative correlation. (It is not necessary to draw the scatter diagrams.)

a	Day		1	2	3	4	5	6
	No. of ripe courgettes		3	5	4	7	9	9

b	Height (cm)	166	156	152	160	159	158	163
	Age (years)	20	31	25	28	22	32	32

c	% seeds germinating in compost 1	51	60	32	70	20	78
	% seeds germinating in compost 2	55	75	70	58	82	50

d	W.p.m. typing test part 1	56	43	49	31	58	60	35	35	44	38
	W.p.m. typing test part 2	58	44	47	37	57	57	30	31	44	43

6 The following table shows the amount of cod landed in the years 1981–86 and percentages of babies vaccinated against measles for the same years.

Cod landed (1000 tonnes)	116	114	112	91	90	77
% vaccinated against measles	54	56	59	63	68	71

a Draw the scatter graph for the above data.
b Can we deduce from this graph that the greater the number of babies vaccinated, the smaller the cod catch will be?
c Explain the high degree of correlation.

It is important to realise that correlation between two sets of data does not mean that the changes in one variable **cause** the changes in the other variable. In other words, being a particular weight does not **cause** you to be a particular height, nor does being a particular height decide what weight you will be.

Although height is linked to weight, there are many other factors which determine your weight.

There is a story that a survey carried out in Sweden paired the number of storks nesting on houses and the number of babies born in the area. The results showed a positive correlation.

We cannot conclude from this that a large number of storks in the area causes a large number of babies to be born, or vice versa.

The reason for the seeming connection is likely to be a third factor, the number of new houses. Storks prefer to nest on new chimneys and new housing estates attract young couples starting families.

When analysing scatter graphs, care should be taken not to jump to conclusions.

It is possible to find a high degree of correlation between variables which are very unlikely to be connected (particularly if only a few points are plotted).

11.3   Line of best fit

If there is a perfect (linear) correlation between two variables, the points plotted will lie on a straight line.

This means that one variable is proportional to the other and an equation of the form $y = mx + c$ can be found to fit the line through the points.

When experimental data is collected, however, there is usually some error in the readings and so the points plotted will not lie exactly along a straight line but close to one.

EXAMPLE 1

An experiment was carried out to measure the height to which a rubber ball bounced after being dropped from various heights.

Height of drop (m)	0.50	0.75	1.00	1.25	1.50	1.75	2.00
Height of bounce (m)	20	25	50	60	75	75	100

Scatter graph:

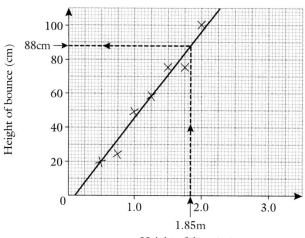

The scatter graph shows that there is a close linear relationship between height of drop and height of bounce. It is not possible to draw a straight line through all the parts so the line which **best fits** all the points is drawn.

All lines of best fit pass through the point $(\bar{x}, \bar{y})$ where $\bar{x}$ and $\bar{y}$ are the means of the first and second sets of data respectively. It is not necessary, for FSMU at Foundation or Intermediate level, to find these means before drawing the line of best fit unless asked to do so.

EXAMPLE 2

Draw the line of best fit for the data of Example 1 and use the line to predict the height of bounce when the drop is 1.85 metres.

Draw, by eye, the line which looks to be the 'line of best fit'.

From the graph:

When height of drop = 1.85 m
height of bounce = 88 cm.

EXERCISE 11.3

1 **a** Draw the scatter diagram for the following data or use your answer to question **1 b** of Exercise 11.1.

Time (minutes)	10	15	20	25	30	35	40
Weight of melting ice block (kg)	3.9	3.5	2.7	1.5	1.2	1.0	0.8

  b Calculate $\bar{x}$ and $\bar{y}$.

  c Draw the line of best fit on your scatter graph.

  d From the graph, find the values of the time when the weight is:
    (i) 2.0 kg   (ii) 3.0 kg

2 **a** The graph below shows the mean weight for a man of medium build and given height.

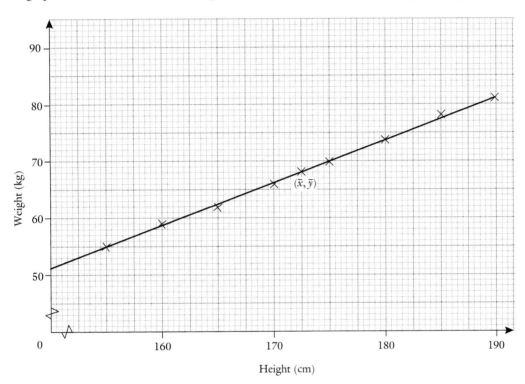

From the graph, find the mean weight for a man whose height is:
(i) 167.5 cm   (ii) 180 cm

  b The corresponding values for a woman's mean weight are:

Height (cm)	142	147	152	157	162	167	172	177
Weight (kg)	46	49	51	54	58	62	65	69

   (i) Plot the scatter graph.
   (ii) Draw the line of best fit.
   (iii) Find the mean weight for a woman of height 165 cm.

3 A mathematics test consisted of two parts: a piece of coursework and a written test. The marks for a class of twelve students are given below:

Student	1	2	3	4	5	6	7	8	9	10	11	12
Coursework (Max 40)	5	7	14	11	20	25	20	29	32	25	32	37
Written test (Max 100)	20	27	32	50	55	62	69	abs	70	90	91	98

a Draw the scatter graph and line of best fit.

b One student completed the coursework, and scored 29 marks, but was absent for the written test.

   Use your line of best fit to determine a written mark for this student.

4 The amount of energy used by a human body just to exist is given below in terms of the body's surface area.

Age (years)	1	3	5	7	9	11	13	15	19
Energy (kcal/h/m^2)	53	51	49	47	45	43	42	42	39

a Draw the scatter graph and find the line of best fit.

b Find the energy required by a person of age 17 years.

c Find the energy required by a person of age 25 years.

d The energy needed by a 25-year-old is actually 37.5 kcal/h/m^2. Explain why the graph gives an inaccurate answer for this age.

5 A county in Southern England recorded information, over a seven year period, showing what pupils did after the age of compulsory education.

The scatter diagram represents these data.

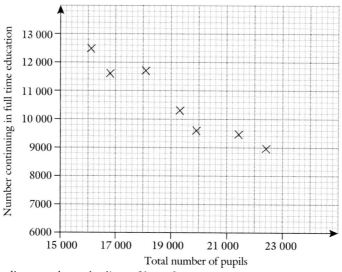

a On the scatter diagram draw the line of best fit.

b Use your line to estimate the number of pupils who would **not** continue in full-time education if the total number of pupils in this age group reached 23 000.

(SEG W95)

12 Interpretation of Statistical Diagrams

12.1 Reading and interpreting diagrams

We have seen in Unit 8 that there are many ways of presenting data in pictorial form. It is clearly necessary to be able to interpret correctly any diagrams given. Examples of pie chart and histogram interpretation are given below.

Interpreting pie charts

The initial interpretation is the fact that the largest portion of a pie chart relates to the largest group, and the smallest portion to the smallest group. However, if any of the data is known, the rest of the data can be calculated.

EXAMPLE 1

The pie chart below shows the number of students in different sections of a college. 220 students are in the Construction department.

a How many students are there in the college?

b How many students are there in Catering?

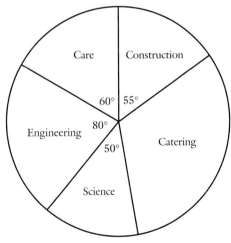

a 55° represents 220 students.

∴ 1° represents $\dfrac{220}{55} = 4$ students.

The complete circle (360°) represents $4 \times 360 = 1440$ students.

∴ There are 1440 students in the college.

b The angle representing Catering is

$360° - (80° + 55° + 60° + 50°) = 115°$.

∴ The number of students in Catering is $4 \times 115 = 460$.

To interpret histograms

EXAMPLE 2

The histogram shows the heights (in metres) of trees in a plantation.
There are 40 trees with heights between 10 m and 12 m.

a How many trees are there with heights between 4 and 6 m?

b How many trees are there in the plantation?

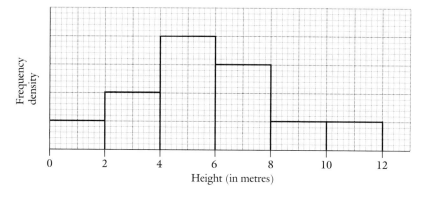

The frequency in a histogram relates to the area of each bar. The 10 to 12 interval has an area of 2 large squares, i.e. each large square represents 20 trees.

a The bar for the 4–6 metre interval has an area of 8 large squares.
There are $8 \times 20 = 160$ trees with heights between 4 and 6 metres.

b The bars have a total area of 24 large squares.
There are $24 \times 20 = 480$ trees in the plantation.

EXAMPLE 3

The histogram represents the speeds of 200 cars along a motorway.

a How many cars are exceeding the 70 mph speed limit?

b What percentage of cars is travelling within 10 mph of the speed limit?

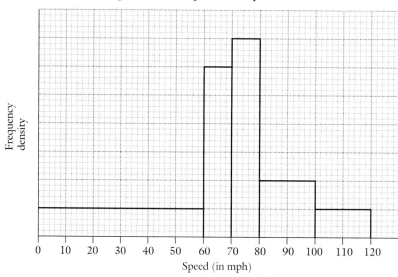

When the histogram has bars of unequal widths, we still use the method described in Example 2.

The total number of large squares is 25.

This represents 200 cars.

$\therefore$ 1 large square represents $\dfrac{200}{25} = 8$ cars.

a The cars exceeding 70 mph are represented by 13 large squares.

   $\therefore$ $13 \times 8 = 104$ cars are exceeding the 70 mph speed limit.

b Cars travelling between 60 and 80 mph are within 10 mph of the speed limit.

   These cars are represented by 13 large squares.

   $\therefore$ There are $13 \times 8 = 104$ cars within 10 mph of the speed limit.

   $\therefore$ The percentage of cars within 10 mph of the speed limit $= \dfrac{104}{200} \times 100 = 52\%$

EXERCISE 12.1

1 The pie chart shows the different drinks sold at lunchtime in college.
720 drinks were sold in total.

Find the number of

a Coke

b orange

c coffee

d chocolate drinks sold during lunchtime.

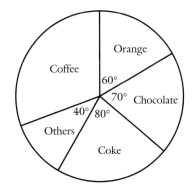

2 The pie chart shows the number of passengers flying from London to Miami
on one afternoon. 1800 passengers in total flew this route on that afternoon.

Find the number flying:

a Virgin

b American Airlines

c British Airways.

The plane used by Virgin is a
Boeing 747 seating 370.
What percentage of the Virgin
seats were occupied?

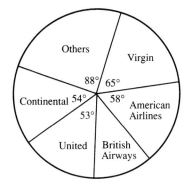

3   A bookshop sold 1080 books, and noted the types of book sold.

Find the number of books
sold which were classed as:

a Thriller

b Hobby

c Travel

d Science.

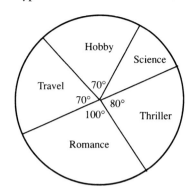

4   The pie chart shows the results of an election
in a constituency. There are 72 000 voters
of whom 80% voted. Which party won,
and what was the winning margin?

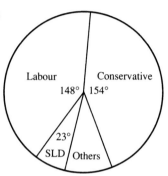

5   The pie chart shows the different petrols sold in one week.

The garage sold 25 000 gallons of diesel.

a How much unleaded petrol
was sold?

b How much lead replacement petrol
was sold?

c What were the total sales?

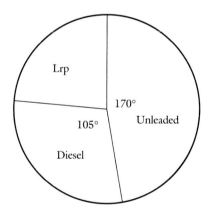

6   The pie chart shows the types of dwelling in which people in a village live.
There are 720 dwellings in the village. By measuring the angles, find how
many are:

a detached houses

b bungalows

c semi-detached houses.

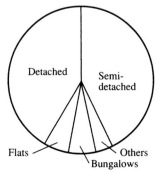

7   The histogram shows the distance travelled by 200 lobsters in one week.
Reconstruct the frequency table.

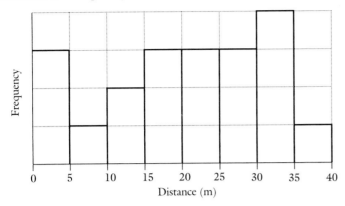

8   At a fête, the number of peas in a jar was guessed. The histogram
represents the guesses made.
Reconstruct the frequency table.

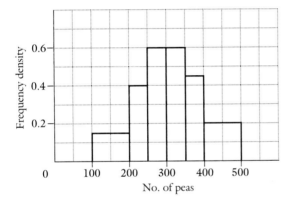

9   The histogram below represents the distribution of ages in a small village.
Find the number of people who live in the village.

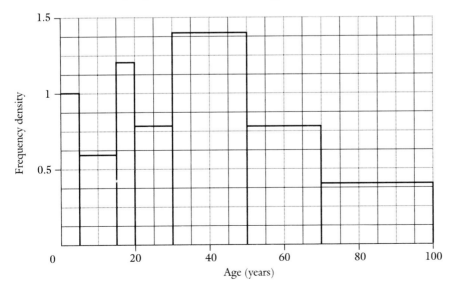

10   The histogram shows the number of cars arriving at a ferry terminal in the last hour before departure. 250 cars arrived in this hour. The latest official check-in time is half an hour before departure.
How many cars arrived after the official check-in time?
What percentage of the cars arrived within 10 minutes of the check-in time?

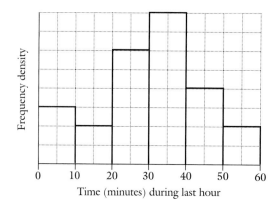

11   The following histogram shows the number of metal rods, of different lengths, made in a factory in one week.
Find the total number of rods.

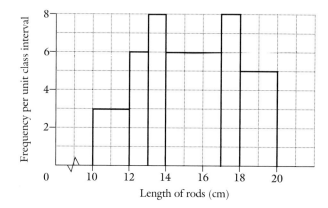

12.2   Drawing inferences from diagrams

When you can readily transfer data to pictorial form, and can convert pictorial representation back into numerical data, you can start to draw inferences from the data given in either form.

Drawing inferences from statistical data is not an exact science. There is rarely a correct, precise answer. If some of the data does not fit the overall pattern, this should be noted. Then, the reasons for the apparent contradiction should be considered.

We can illustrate this by examples.

EXAMPLE 1

The number of people each day visiting an open-air swimming pool in August were:

341, 352, 347, 355, 361, 341, 344, 352, 344, 360, 371, 347, 329, 351, 621, 357, 348, 359, 354, 372,....

What can be inferred from this data?

The 15 August figure of 621 is clearly exceptional and should be easily identified.

The reasons why this figure is exceptional would be unlikely to be found without further questioning – for example:

(i) was an alternative swimming facility closed for the day?
(ii) was the weather on the 15th substantially hotter than on the other days?
(iii) does the data refer to a country with a Bank Holiday on 15 August (e.g. France)?

These are all possible reasons and you may be able to suggest others.

Some data gives results from which suitable inferences can be fairly quickly drawn.

EXAMPLE 2

A men's clothes shop has two branches: one in a city centre, and one at an out-of-town shopping complex. The weekly sales of the two shops over a period of three months are given below:

Date (week commencing)		December 7	14	21	28	January 5	12	19	26	February 2	9	16	23	March 2	9
Sales in thousands of pounds	City shop	18	21	23	14	29	31	28	7	5	8	6	4	5	8
	Out-of-town shop	22	27	52	15	4	5	6	8	9	8	10	9	7	8

What can be inferred from this data?

From the above data, the following inferences could be made:

(i) Sales immediately prior to Christmas are higher than at other times in the three-month period.
(ii) Sales in the out-of-town complex are generally higher than those in the city shop.
(iii) The city shop had a 'sale' during the first three weeks of January.
(iv) During the 'sale' the trade at the out-of-town shop was reduced.

In reality, the firm could well investigate the types of garments sold over the years to decide whether or not to promote the same articles in both shops at the same or different times. Computerisation of sales enables shops to keep far better checks on stock sold. This enables them to react more quickly to consumer demand and to supply each individual shop with the goods which its specific customers require.

EXERCISE 12.2

You may find it helpful to study Unit 9 (Averages) before doing this exercise.

1   In an Ian Fleming novel the number of words per sentence in the first 100 sentences are as follows.

Number of words	0–4	5–9	10–14	15–19	20–24	25–29	30–34	35–39
Number of sentences	14	37	28	12	5	1	1	2

a   Copy the axes and draw a frequency polygon to illustrate this information.

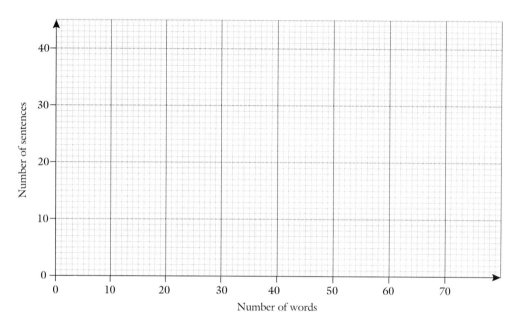

The frequency polygon of the number of words per sentence in the first 100 sentences of a novel by a different author is drawn below.

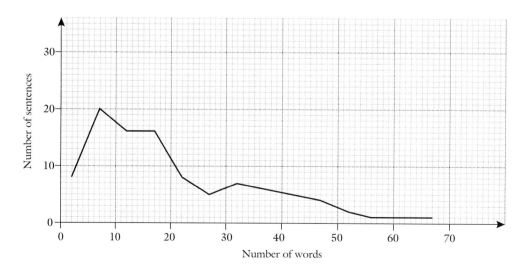

b   Use the frequency polygons to compare the sentence length of the two novels.

(SEG Sp94)

2 A plant has two varieties **A** and **B**.
A study was made of the height of the two varieties.
The histograms represent the frequency distributions obtained.

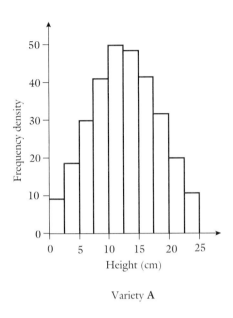

Variety **A**

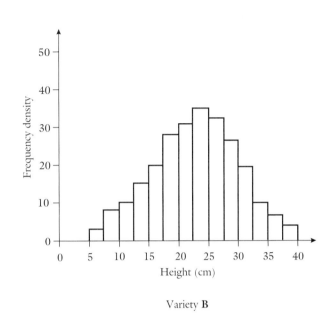

Variety **B**

a Name the type of distribution shown by the height of these plants.

b Make **two** statistical comments about the height distributions of these plants.

(SEG W94)

3 The height, to the nearest inch, of 500 adult females is given in the following table.

Height	58	59	60	61	62	63	64	65	66	67	68	69	70	71	72
Frequency	1	0	4	3	9	32	67	84	102	135	41	18	3	0	1

a Complete the grouped frequency table.

Height (inches)	Class mid-point	Frequency
54.5–59.5	57	1
59.5–64.5		
64.5–69.5		
69.5–74.5	72	

b Use your grouped frequency distribution from **a** to calculate an estimate of the mean height of the 500 females.

c The frequency polygon showing the distribution for the height of 500 adult males is shown below. Copy it and on the same axes, draw the frequency polygon for the grouped distribution of the female height.

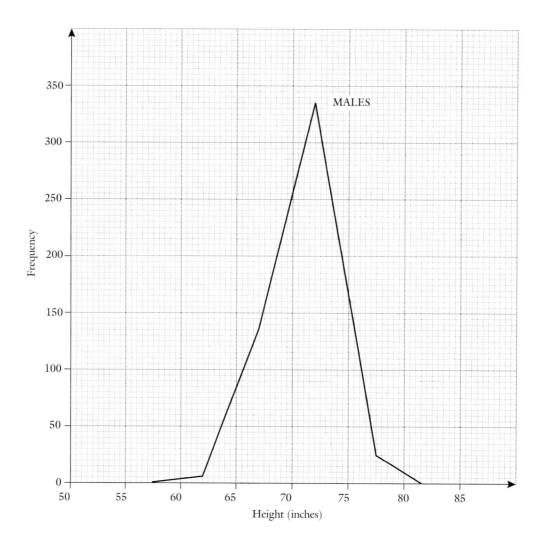

d Use the frequency polygons to compare the means and dispersions of the two distributions.

(SEG Sp94)

12.3   Dangers of visual misrepresentation

Statistics can very easily be presented in a form which, although correct, is misleading. The simplest way in which this occurs is by the use of the 'false origin'.

For example, when a rail line is electrified, the number of cars per day on the parallel road reduces from 38 500 per day to 38 200.

A correct bar chart would show:

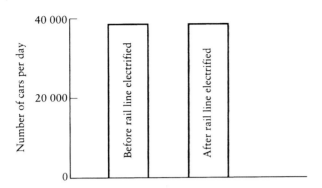

From this, readers would draw the conclusion that there has been little change.

Carefully deleting most of the vertical scale, and starting at **38 000** produces the bar chart below:

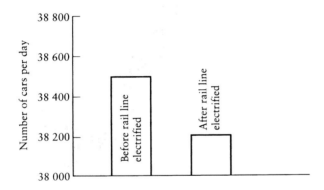

To a casual reader, the conclusion would be quite clear: there has been a significant reduction in the number of cars using the road.

Although this would rarely be spotted, the scale can be more subtly altered as below:

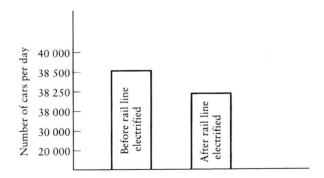

This over-expansion of the relevant part of the vertical scale can be justified as giving prominence to that part of the graph in which we are interested.

However, its effect is to mislead readers.

If a section of a scale is to be deleted, it should be clearly identified – usually with a squiggle on the axis as shown on the right:

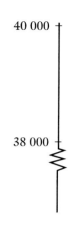

Bars of different widths can also mislead:

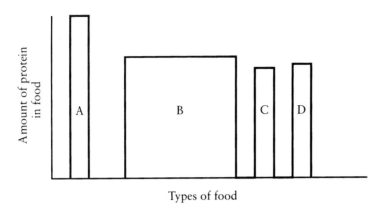

Although B does not have the highest amount, a casual readers' eye is drawn to it as its 'area' is much greater than any other bar.

The phrase 'amount of protein in food' is also open to misrepresentation, as it does not specify how much of food A is compared with how much of food B. Are they proportions, or amount per penny, or amount per gram, or something else?

False impressions can also be given by careful selection of which figures to show. In a time-series, you have to have a start and end date. In the presentation of information demanding an increase in wages, salaries or subsidies, it is to be expected that the data will start at a peak year. In other cases the data will stop at a favourable moment.

EXAMPLE

The diagram shows the Japanese share index for 1981–90. The diagram on the right shows the same index stopping in Summer 1987, when it would show an unbroken climb.

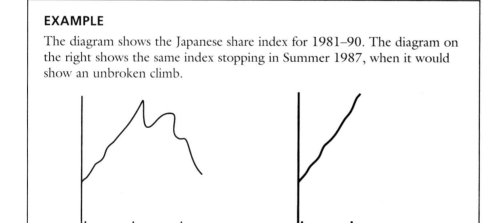

All diagrams should:

(i) be clearly labelled and titled

(ii) have the scales clearly identified

(iii) have the units given

(iv) should be drawn as outlined in Unit 8.

Any diagram which is not so drawn may, intentionally or not, mislead.

EXERCISE 12.3

Criticise the diagrammatical representations shown in the following questions:

1

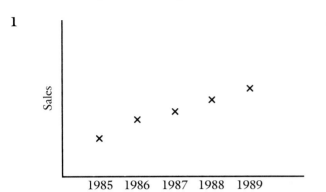

5

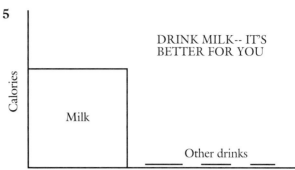

2

6

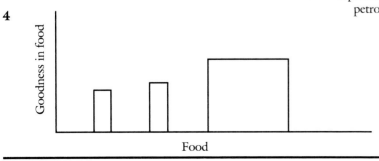

3

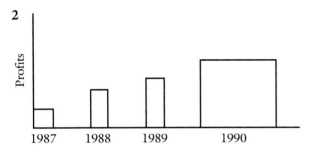

7

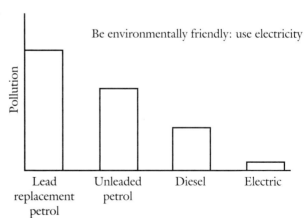

4

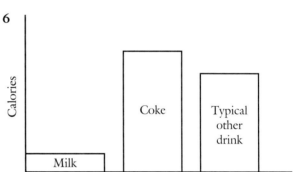

12.4 Interpretation of statistical information

All statistical information needs to be considered carefully before any inference can be made. It is rare for a simple set of statistics to prove anything.

An advertisement states '60% of dog owners, who expressed a preference, liked Dogamix'. It could be that out of 1000 owners, only 5 showed a preference; i.e. 3 out of 1000 liked it and 2 out of 1000 disliked it.

If a packet states average net weight of 250 grams, about half of the packets bought will have weight less than 250 grams, and half will have weight over 250 grams. Hence it is difficult for shoppers to tell whether they are being sold too little.

Professional statisticians are often involved in work of this nature. They are often employed to obtain, and present, the information in as favourable a way as possible for their employer. Detailed work in this area is beyond the scope of this course, but readers should always approach statistical information with care.

13 Probability

13.1 Introduction

Probability is a measure of how likely something is to occur. What occurs is an **outcome**.

This likelihood or probability can be represented on a sliding scale from the probability when an event is certain to occur to the probability when the event cannot occur.

Suppose twenty children in class 2X are studied. It is found that 19 do not wear glasses, and 1 does wear glasses. 11 of the 20 children are girls and 9 are boys.

Suppose a child is picked at random. The likelihood of the following events occurring fits into the following pattern.

Event	Likelihood
The child studied is in class 2X	Certainty
The child does not wear glasses	High probability
The child is a girl	Just over half
The child is a boy	Just under half
The child wears glasses	Highly unlikely
The child's age is over 70	Impossible

Since 19 out of 20 children do not wear glasses, the likelihood that a child selected at random does not wear glasses is 19 out of 20, or $\frac{19}{20}$. Therefore, the probability that a child selected at random is not wearing glasses is $\frac{19}{20}$.

If all 20 children are wearing shoes, the likelihood that a child selected at random is wearing shoes is 20 out of 20, i.e. $\frac{20}{20}$, which is 1. Therefore, the probability is 1.

None of the children have green hair. The probability that a child selected at random has green hair is 0 out of 20, i.e. $\frac{0}{20}$, which is 0.

If an event is bound to occur, then its probability is 1.

If an event cannot occur, then its probability is 0.

The probability of a child not wearing glasses is $\frac{19}{20}$.

The probability of a child wearing glasses is $\frac{1}{20}$.

The probability of a child wearing glasses or not wearing glasses is 1 (i.e. a certainty).

Note that $\frac{19}{20}$ (not glasses) $+ \frac{1}{20}$ (glasses) $= \frac{20}{20} = 1$ (certainty).

From this we can work out that:

(i) **The total probability for all possible outcomes is 1.**

(ii) **Probability of event happening $= 1 -$ Probability of it not happening**

The events in the initial example occur with the following probabilities:

Event	Likelihood	Probability
The child studied is in class 2X	Certainty	1
The child does not wear glasses	High probability	$\frac{19}{20}$
The child is a girl	Just over half	$\frac{11}{20}$
The child is a boy	Just under half	$\frac{9}{20}$
The child wears glasses	Highly unlikely	$\frac{1}{20}$
The child's age is over 70	Impossible	0

13.2   Probability from theory and experiment

Probability can be found either by theory or by experiment.

The theory method relies on logical thought; the experimental method relies on the result of repetition of the event producing results which are taken to be typical.

EXAMPLE

Find the probability of getting a head when tossing an unbiased coin.

Theory method
'Unbiased' means both events (head, tail) are equally likely to occur.

Heads and tails are the only outcomes,
∴   Probability of head + Probability of tail = 1
(since total of all possible outcomes is 1)

Probability of head = Probability of tail
∴   Probability of tail = $\frac{1}{2}$.

Experimental method
Toss an unbiased coin 100 times, and count the number of tails you obtain. You might obtain 49 tails.
∴   The experimental probability of getting a tail is $\frac{49}{100}$.

If the event is repeated many times, it is usual for the experimental probability to be close to the theoretical probability, but not to be exactly the same.

For the probability of a person being right-handed, there is no theory to use. This would be found by experiment. For example, ask 100 people and see how many are right-handed. The more people you ask, the more likely it is that your probability is accurate, but it must be appreciated that the probabilities found by experiment are not exact. The result may also be different if the experiment is repeated with another group.

Similarly, if the probability of a car being red is $\frac{1}{4}$, the expected number of red cars in a car park containing 80 cars is

$$\frac{1}{4} \times 80 = 20$$

1 The probability that a street light is lit is $\frac{3}{8}$.
   What is the probability that it is not lit?

2 You toss a coin, marked with a head on both sides.
   What is the probability of getting a tail?

3 The probability of a plane being late is $\frac{2}{3}$.
   What is the probability of a plane not being late?

4 Jane has a bag containing 6 oranges and 4 apples. She selects a fruit from
   the bag at random.
   What is the probability that it is a pear?

5 Probabilities can be estimated **either** by making a subjective estimate **or**
   by making use of statistical evidence.
   State the method which would be used in the following cases.

   a The probability that more women than men will read the news on
      television.
   b The probability that man will live on the moon in the year 2050.
   c The probability that the next vehicle passing a college will be a motor
      cycle.
   d The probability that the next book issued at a library will be a novel.
   e The probability that there will be a cure for deafness within 10 years.
   (SEG S95)

6 In an experiment a drawing pin is thrown. The number of times it lands
   point up is recorded.

   The drawing pin is thrown 10 times and lands point up 6 times.

   a From these data estimate the probability that it lands point up.

   The drawing pin is thrown 100 times and lands point up 57 times.

   b From these data estimate the probability that it lands point up.

   c Give a reason why the second answer is a more reliable estimate.
   (SEG S94)

Solutions to Exercises

Note. Solutions are given for exercises with a 'closed' numerical answer. They are not provided for 'open-ended' questions and for some questions leading purely to an illustration (e.g. a pie-chart or a graph).
Numerical answers are generally given correct to 3 significant figures.

Exercise 1.1

1 **a** 5 tens   **b** 7 hundreds   **c** 3 thousands
   d 4 units   **e** 6 tenths   **f** 2 hundredths
2 **a** 10, 17, 21, 45, 54, 86
   b 14, 23, 32, 104, 203, 230
   c 6, 60, 61, 600, 601, 610
   d 11, 99, 101, 110, 999, 1001
   e 297, 300, 399, 400, 420
3 7510   4 2369
5 **a** 430   **d** 132   **g** 516
   b 16 700   **e** 590   **h** 7030
   c 20 000   **f** 935.8   **i** 130
6 **a**   32   **d** 40.3   **g** 0.0027
   b   3   **e** 0.65   **h** 0.051
   c   40   **f** 0.065   **i** 0.0002
7 **a** 100   **b** 400   **c** 4100   **d** 6200

Exercise 1.2

1 41.9   2 180.3   3 133.97   4 16.7   5 16.71
6 51.75   7 169.4   8 702.36   9 1770   10 7.8
11 72.31   12 47.86

Exercise 1.3

1 26   2 36   3 5   4 1   5 1   6 7
7 0   8 66   9 133   10 2   11 £4.96   12 4
13 £78.01   14 **a** 988   **b** £4755
15 **a** 2 h 7 min (127 min) **b** £19.05
16 7 packs, £1.35   17 **a** 47.5 miles **b** £12.92
18 £1183.80   19 £160.50
20 **a** 7.39 kg **b** 135   21 **a** 9 **b** 14.6 cm

Exercise 1.4

1 2 **b** −4 **c** −10 **d** 4 **e** 6 **f** 40 **g** 0 **h** −49
2 **a** −1 **b** 17 **c** −13 **d** 20 **e** −3 **f** 0 **g** 50 **h** 36

Exercise 1.5

1 **a** −18 **b** −3 **c** 28 **d** $1\frac{3}{4}$ **e** −15
   f −27 **g** 8 **h** 36 **i** −1 **j** −10

Exercise 1.6

1 1, 4, 9, 16, 25, 36, 49, 64, 81, 100
2 1, 8, 27, 64, 125, 216
3 **a** 5 **b** 11 **c** 25 **d** 15 **e** 14
4 **a** (i) 3 (ii) −4 (iii) 9
   b 2 **c** (i) 3 (ii) 5
5 1 + 3 + 5 + 7 + 9
6 8 7 27 8 **a** 2 **b** 3 **c** 6 **d** $\frac{1}{2}$ **e** −4
9 3 10 100

Exercise 2.1

1 9.74 2 0.36 3 147.5 4 29
5 0.53 6 4.20 7 1250 8 0.004
9 270 10 460.0

Exercise 2.2

1 **a** 3500 **b** 240 **c** 1200 **d** 300 **e** 250
   f 10 **g** 40 **h** 5
2 **a** 9 **b** 0.06 **c** 24
3 Incorrect calculations are **c, d, e, f, h, i.**

Exercise 2.3

1 52p 2 6 tins 3 5 4 **a** 27p **b** 3p
5 2m 6 25p 7 £10.91 8 £13.86
9 £1458.90 10 26

Exercise 2.4

1 11.8 2 73 700 3 2.70 4 0.173 5 88.9
6 29.0 7 21.4 8 26.0 9 0.786 10 882.65
11 0.829 12 11 13 3.3

Exercise 3.1

1 **a** 8 **b** 8 **c** 18 **d** 9 **e** 7 **f** 9
2 **a** $\frac{2}{3}$ **b** $\frac{3}{4}$ **c** $\frac{2}{9}$ **d** $\frac{1}{3}$ **e** $\frac{3}{4}$ **f** $\frac{2}{3}$ **g** $\frac{6}{13}$ **h** $\frac{2}{3}$
3 $\frac{2}{3}$ 4 **a** $\frac{9}{16}$ **b** $\frac{3}{16}$ 5 $\frac{3}{16}$ 6 $\frac{1}{4}$ 7 $\frac{4}{7}$
8 **a** $\frac{20}{9}$ **b** $\frac{31}{6}$ **c** $\frac{59}{12}$ **d** $\frac{111}{10}$ **e** $\frac{35}{4}$ **f** $\frac{38}{3}$ **g** $\frac{143}{20}$ **h** $\frac{143}{7}$
9 **a** $1\frac{2}{7}$ **b** $7\frac{1}{5}$ **c** $3\frac{1}{3}$ **d** $7\frac{5}{8}$ **e** $13\frac{10}{11}$ **f** $10\frac{7}{12}$
   g $8\frac{4}{5}$ **h** $23\frac{1}{2}$

Exercise 3.2

1 $1\frac{3}{20}$ 2 $2\frac{1}{24}$ 3 $\frac{2}{3}$ 4 $\frac{3}{4}$ 5 $\frac{3}{20}$
6 **a** 6 **b** (i) $\frac{1}{4}$in (ii) $\frac{11}{16}$in 7 $\frac{1}{12}$
8 $\frac{2}{15}$ 9 **a** $\frac{1}{12}$ **b** 3
10 **a** $\frac{3}{25}$ **b** 400 mg aspirin, 40 mg ascorbic acid, 60 mg caffeine
11 $3\frac{5}{12}$ 12 7 in 13 $\frac{3}{16}$ in 14 $1\frac{7}{8}$ in

Exercise 3.3

1 **a** $\frac{8}{12} = \frac{2}{3}$, 0.6̇, **b** $\frac{1}{4}$, 0.25 **c** $\frac{5}{8}$, 0.625
2 **a** 0.1 **b** 0.5 **c** 0.75 **d** 1.45
   e 4.84 **f** 2.83̇ **g** 7.4̇ **h** 3.1̇42857̇
3 **a** $\frac{1}{2}$ **b** $\frac{1}{4}$ **c** $\frac{2}{3}$ **d** $1\frac{1}{3}$ **e** $1\frac{3}{10}$
   f $2\frac{4}{5}$ **g** $3\frac{3}{5}$ **h** $2\frac{3}{20}$ **i** $\frac{1}{8}$ **j** $\frac{3}{8}$
5 **a** Insert d.p. into numerator 1 place from right.
   b Multiply numerator by 2 and change to tenths.
   c Insert d.p. into numerator 2 places from right.
   d Change to tenths then divide by 2.
   e Multiply numerator by 4 and change to hundredths.
6 34 7 **a** (i) $1\frac{2}{3}$ (ii) 1.67 **b** 80p
8 **a** 20 sheets **b** (i) $20\frac{5}{6}$ (ii) 20.83
9 (i) $4\frac{2}{3}$ hours (ii) 4.67 hours
10 (i) $3\frac{1}{5}$ hours (ii) 3.20 hours 11 (i) $1\frac{3}{4}$ (ii) 1.75

Exercise 4.1

1 a 20% b $12\frac{1}{2}$% c 70% d 65% e $66\frac{2}{3}$% f 36%
  g 175% h 250%

2 a $\frac{3}{5}$ b $\frac{1}{4}$ c $\frac{1}{10}$ d $\frac{17}{20}$ e $\frac{3}{20}$ f $1\frac{3}{20}$ g $\frac{3}{8}$ h $\frac{1}{3}$

3 a   0.75   75%
  b $\frac{1}{2}$ 50%
  c   0.125 $12\frac{1}{2}$%
  d $\frac{1}{3}$ $0.\dot{3}$
  e $\frac{3}{8}$ $37\frac{1}{2}$%
  f   0.7   70%
  g $\frac{7}{20}$ 0.35
  h $\frac{2}{3}$ $66\frac{2}{3}$%
  i   0.6   60%
  j $\frac{5}{8}$ 0.625

Exercise 4.2

1 a £1.38 b £318.62 c £22.12 d £13.91 e £5.17
  f £8.73 g £97.20 h £44.20

Exercise 4.3

1 a £10.12 b £4.07 c £5.50 d £10.79 e £10.04
2 a £108.18 b £1.05 c £792.00 d £233.75
  e £156.00 f £21.69

Exercise 4.4

1 a £24.00 b £8.40 c £14.08 d 36p e £10.39
2 a £49.77 b 59p c £51.20 d £23.76 e £59.43
  f £1.09

Exercise 4.5

1 a 80.0% b 3.45% c 175% d 50% e 54.5% f 130%
2 a +23.1% b +20.0% c +375% d −20.0%
  e −45.5% f +464%

Exercise 4.6

1 a 18% 2 a 260 b 78 3 12%
4 a 20 000 b 7500 c 36.2%
5 a 406 b 146 6 £2.64 7 £3.43
8 £1335.70 9 616 10 £2240

Exercise 5.1

1 Discrete   2 Continuous   3 Continuous
4 Continuous   5 Qualitative   4 Continuous
7 Continuous   8 Qualitative   9 Discrete
10 Qualitative   11 Discrete   12 Qualitative

Exercise 6.1

1 a total population of the town
  b people asked at a particular time in a particular shopping street
  c Specify the intended catchment area of the new shopping centre. Use the electoral role for this area to select a random sample.

2 a all the students in the College
  b She only asks people in the common room, and she only asks girls.
  c Find students' enrolment numbers, and select 50 random numbers in the range of numbers given.
3 a all cars in the UK
  b The police only check cars between 5pm and 6pm.
  c Sample at different times of day on different days of the week on different roads.
4 a the soil in his garden
  b only one sample. Soil beyond the range of his throw cannot be sampled.
  c Draw a plan of the garden and divide into numbered squares. Use random sampling to choose several squares and take samples from these areas.
5 a all teenagers
  b Carol only asked people entering a tobacconist's. The sample is biased towards smokers. Carol should only ask teenagers.
  c Obtain a list of pupils in the school; select a random sample from the list.
6 a any homes in the telephone area
  b The salesgirl only asks those with a telephone. The sample includes only those who are at home when the salesgirl rings.
  c The salesgirl needs to know the number of homes in the area. She needs to use the rating register (used in water rates) to identify homes; use their reference numbers to obtain a random sample.
7 a all cars owned by people in an area
  b Peter only asked people at home during one afternoon. The responses came from a small part of the town.
  c Use the DVLC in Swansea to produce a list of cars registered as being owned by people in the required area. Use random numbers to obtain a sample of the required size.
8 a all people who travel to work in John's area
  b People at the station would be biased towards those using a train to go to work.
  c Use the electoral role and ask a random sample from this how they travel to work. If they do not work, delete them from the sample. Continue with a random selection until you have a suitable number who do work.

Exercise 6.3

1 a Analyse rainfall for the given period over several years.
  b Devise a questionnaire, survey, or experiment.

Exercise 7.1

(Frequencies only are given.)
1 8, 7, 5, 5, 4, 3, 1, 2, 1, 1, 1, 0, 1, 0, 1
2 1, 3, 5, 10, 19, 6, 6
3 18, 17, 13, 7, 4, 2, 1, 0, 1
4 4, 8, 9, 8, 9, 6, 2, 2, 2
5 a 1, 4, 4, 4, 8, 5, 4, 3, 5, 1, 1
  b 1, 8, 15, 8, 6, 2

6 1, 14, 23, 19, 6
7 5, 6, 2, 8, 10, 12, 3, 4
8 9, 11, 8, 11, 5, 8, 12, 11, 5
9 1, 7, 15, 16, 9, 2
10 31, 13, 15, 1
11 1, 0, 5, 16, 13, 4, 1

Exercise 8.1

1 **a** East Lynne **b** 50 **c** 410
4 **a** 175 **b** 8.3% **c** 4 : 1
7 **a** Soft white baps **b** 6 **c** 61.9(62)%
9 **a** 20 **b** Evensong **c** 110
11 **a** Plymouth–Roscoff **b** $3\frac{3}{4}$ hours **c** $2\frac{1}{2}$ hours
13 **a** 5 **b** Train **c** 150

Exercise 8.2

1 **a** Oak **b** Oak 80; Elm 10; Chestnut 20;
   Beech 40; Conifer 70; Cedar 25 **c** 245
6 **a** A; Adverts on hoarding B are aimed at more
   prosperous suburban inhabitants.
   b 2 **c** Football club and health education
10 **a** £30 000 **b** £27 500 **c** £2500 **d** Central

Exercise 8.3

1 **a** 160 **b** 120 **c** 220 **d** 140
6 **a** 40° **b** 25 **c** 50 **d** $\frac{5}{18}$
10 54 762, 5952, 85 714
12 **a** 298 **b** 46 **c** 306
16 **a** 325 **b** 290 **c** 210, 87.8%
18 **a** 72.2% **b** 458 300 acres

Exercise 8.5

5 **a** 1 040 000 **b** 1927, 1951, 1974 **c** 1940
6 **a** Tiredness as the week goes by **b** 226
7 **a** £180 **b** £209
   c Line beginning to curve upwards slightly **d** 1990
8 **b** Summer is peak time, but numbers of passengers
   could be declining, whereas winter numbers seem to
   be increasing.
10 **a** September; new registration letter
   b 15 **c** 171 **d** 29

Exercise 8.7

2 **b** A low absence rate is more likely on Mondays and a
   high one on Fridays.
3 **b** Book 1: Science fiction.
   Book 2: Child's story.
   Book 1 has a greater number of long words and Book
   2 has a greater number of short words.
4 There has been a move to 'Traditional Style' from
   'Modern Design'.
6 There are more women than men over the age
   of 45.
7 **b** The experimental results are very close to the expected
   results.
8 There were less faults on the new model.

Exercise 8.8

```
1   7 | 2 2            2   9 | 2
    6 | 1 1 5 5 8          8 | 1 4
    5 | 3 4 7             7 | 0 1 1 4 6
    4 | 1 1 1 5 8 9       6 | 1 2 3 3 5 7
    3 | 1 3 7 9           5 | 4 8 9
                         4 | 1 9 9

3   9 | 1 2            4   7 | 1
    8 | 2 5 9             6 | 2 8
    7 | 1 4 5             5 | 2 3 3
    6 | 1 4               4 | 1 7 7 8 8
    5 | 2 7               3 | 5 6 8 9
    4 | 8 9               2 | 1 7 7 8
```

Exercise 8.9

```
1        Monday   |    Saturday
                  | 7 | 1
                  | 6 | 7 9
                  | 5 | 1 1 1 1 3 6
                4 | 4 | 1 2 2 7 9 9
        9 4 4 3 2 1 | 3 | 1 1 9
      0 0 8 7 5 4 1 1 | 2 | 5 8 9
    9 9 7 7 5 5 5 3 1 | 1 | 1
                  | 0 | 3 4
```

```
2       Wednesday  |    Saturday
                1 | 3 | 0 0 1
            8 3 1 | 2 | 1 4 5 7 8 8 9 9
      9 7 6 5 5 2 1 | 1 | 2 5 7 9
            9 8 6 | 0 |
```

```
3        Portugal  |    England
          1 2 2 2 | 9 |
    2 8 9 3 4 7 5 7 1 | 8 | 1
        1 9 2 4 9 | 7 | 1
              9 7 | 6 |
            2 2 4 | 5 | 8
          2 1 1 9 | 4 | 7 1 5 7 8 9
                1 | 3 | 3 2 7 8 9 7 5 6 3 9 9
                  | 2 | 3 1 8 9
            5 1 9 | 1 | 1 7 5 9
                  | 0 | 1 2 1
```

Unordered Steam-and-leaf:

Portugal		England	
LQ = 4.9	$\bar{x} = 6.616$	LQ = 2.1	$\bar{x} = 3.394$
Median = 7.4	$s = 2.422$	Median = 3.6	$s = 1.810$
UQ = 8.7	IQR = 3.8	UQ = 4.5	IQR = 2.4
Range = 9.2 − 1.1 = 8.1		Range = 8.1 − 0.1 = 8	

Portugal gets about twice the hours of sunshine that England gets. Also variability of sunshine hours for Portugal is about $1\frac{1}{2}$ times that of England. (means: 6.616; 3.394 and medians: 7.4; 3.6).

4

		Male		Female
	2 1	5		
	7 7 2	4		
8 8 5 4 1		3	1 1 2 4 8	
9 8 5 5 1 1		2	1 1 1 4 4 5	
9 7 7 5		1	1 2 2 5 7 7 7 9 9	
8		0	8	

Exercise 9.1

1 £130   2 68.27 mph   3 38.58
4 The ages are given to the last completed year. A person aged 17 is between 17 years 0 days and 17 years 364 days old. Therefore, the true mean is probably above $28\frac{3}{4}$, hence John is probably correct.
5 1.848 m   6 104.8   7 a 1.29 m   b Yes
8 a 1125   b 75 w.p.m.
9 a £496.14   b £5.04   10 13.5
11 a 164 lb   b 32.8 lb   c $13\frac{2}{3}$ (13.7) lb per week
12 a 21.9°C   b (i) 18.75°C   (ii) 24°C
13 a 51.78   b 57.89 m
 c Considerable improvement (over 6 m)
14 Average is 40.9 mpg. Therefore, taking values to 2 s.f., he is justified.
15 6 min 5 seconds

Exercise 9.2

1 a 4   b 4   2 Blue   3 110–114
4 a 12.5   b 11   5 6.35 kg   6 1
7 a James and Linda   b 5 ft 2$\frac{1}{2}$ in   c Mode
8 a 50p   b 44p
9 a £10 000   b £10 000   c £16 800
 d Mode or median
10 a 9.00–11.00   b 70–75
11 a 18   b 16   12 St Lucia
13 a £112   b £112   14 a 4   b 2.5
15 4:55 (4 min 55 sec)

Exercise 9.3

1 Mode   2 Median   3 Mode   4 Median   5 Mean
6 Mean or median   7 Median   8 Mean   9 Mean
10 Any, but the mean is preferable.

Exercise 9.4

1 6   2 26.60 min (= 26 min 36 s)   3 7.79 words
4 a 1.43   b 1   c 0   d Median
5 a 4.17   b 4   6 a 7.5   b 7.06
7 a 3.5   b 3.28   8 a 1   b 1
9 a 4.08   b 4   c Median
10 a 2.62   b 3   11 a 2.94   b 3
12 a 0.52   b 0   c Mean
13 a 19   b 19.3
 c Close match between mean and median

Exercise 9.5

1 20 in   2 a 12.6 min   b 10 min
3 a (i) 31.8   (ii) 36.3
 b (i) 50   (ii) 25   4 34   5 8.9
6 a 38.2, 41   b 56, 22   c Beach 2. It has a slightly higher average number of shells and is more consistent.
7 a £16.10, £2.55, £16.82, £2.60
 Similar ranges, but Club 2 has a higher mean.
 b £15.80, £16.95
 c Club 1: distribution is possibly skewed since median < mean; Club 2: distribution is negatively skewed since median > mean.
8 a 122.3, 82.5   b (i) 30   (ii) 30
 c Blood pressure high on day 8, both pressures below normal on day 10.
9 a 55.8 cm, 55.5 cm   b 51.0 cm, 114 cm
 c The first archer, who is more consistent
10 a 50.3, 48.7   b 4.47, 2.94   c Machine 3 has a mean close to 50 but is variable.
 Machine 5 is producing ball bearings which are consistently below size and should be adjusted.

Exercise 10.1

1 a 32   b 28   c 35   d 7   e 8
2 a 73 cm   b 66 cm   c 79 cm   d 13 cm
 e (i) 4   (ii) 55
3 a £12 700   b £11 200   c £13 800   d £2600
 e £14 100
4 a 58   b 44   c 71   d 27   e (i) 84   (ii) 83
5 b (i) 190 g   (ii) 74   c (i) 34 kg   (ii) 21 kg
6 a 30, 90, 130, 150, 160
 c 1 min 34 s (1.57 min)
 d 20

Exercise 10.2

7

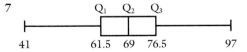

8

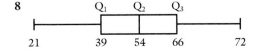

Exercise 11.2

1 a Weak positive   b High negative   c None
2 Weak positive   3 Quite high negative
4 Quite high positive
5 a Positive   b None   c Negative   d Positive
6 b No
  c Over a period of time both variables have changed, but this is a coincidence. The smaller cod catches are probably due to depleted fish stocks. The percentage vaccinated refers to an independent event.

Exercise 11.3

The equations of the lines are given to enable students to check the position of their lines. Answers given are by calculation, answers from graphs should be correct to two s.f.
1 a $\bar{x} = 25$, $\bar{y} = 2.09$
  c $y = -0.113x + 4.91$
  d (i) 25.8 min   (ii) 16.9 min
2 a (i) 64 kg   (ii) 73.8 kg
  b (ii) $y = -48.8 + 0.662x$ (iii) 60.4 kg
3 a $y = 13.1 + 2.28x$
  b 79 marks
4 a $y = 52.9 - 0.781x$
  b 39.6 kcal/hour/m^2
  c 33.3 kcal/hour/m^2
  d The relationship is not linear outside the given range.
5 b 14 500

Exercise 12.1

1 a 160   b 120   c 220   d 140
2 a 325   b 290   c 210, 87.8%
3 a 240   b 210   c 210   d 120
4 Conservative by 960 votes
5 85 714
6 a 298   b 46   c 306

7 Distance (m)	Number of lobsters
0–5	30
5–10	10
10–15	20
15–20	30
20–25	30
25–30	30
30–35	40
35–40	10

8 Number of peas	Frequency
100–199	15
200–249	20
250–299	30
300–349	30
350–399	22
400–499	20

9 81   10 140, 56%   11 56

Exercise 12.2

1 b For example:
  The second author generally uses much longer sentences than the first.
  $\frac{4}{5}$ of Fleming's sentences contain less than 15 words, whereass less than half the second author's do.
  14% of the second author's sentences are longer than any sentence of Fleming's.
  i.e. the sentences in the second novel are longer on average and have a larger variation in length.
2 a Normal distribution.
  b Variety B is generally taller than variety A and has a greater variation in height.
3 a 1, 115, 380, 4   b 65.87 inches
  d The mean height of females is lower than that of males and the standard deviation is less, i.e. females are generally shorter than men and there is less variation in their heights.

Exercise 12.3

1 Missed: 1988, 1989. No scale on sales.
2 Different widths of bars. Non-linear vertical scale.
3 Vertical scale does not start at origin. No time interval on horizontal scale.
4 Different width bars. No vertical scale. No indication of meaning of 'goodness'. Food not identified.
5 Other drinks not identified. Different width bars. No vertical scale. Probably picked one constituent of milk not in other drinks.
6 No vertical scale. Different width bars. 'Typical other drink' meaningless.
7 No vertical scale. 'Pollution' not defined.

Exercise 13.1

1 $\frac{5}{8}$ 2 0   3 $\frac{1}{3}$ 4 0
5 a Statistical evidence   b Subjective estimate
  c Statistical evidence   d Statistical evidence
  e Subjective estimate
6 a $\frac{3}{5}$ or 0.6   b $\frac{57}{100}$ or 0.57   c greater amount of data used.

Index